Selections from *Astronomia Nova*

SELECTIONS FROM

Kepler's *Astronomia Nova*

A Science Classics Module
for Humanities Studies

Selected, translated, and annotated by
William H. Donahue

Manufactured in the United States of America.

Published by Green Lion Press
www.greenlion.com.

Cover illustration: Diagram of Mars's path, according to the Ptolemaic model, from 1580 to 1596. *Astronomia Nova* Chapter 1 (copied from Kepler's original engraving by W. H. Donahue).

Cataloging-in-Publication Data:

William H. Donahue
Selections from Kepler's Astronomia Nova: A Science Classics Module for Humanities Studies

Excerpts of text of Johannes Kepler
Translation, notes and commentary by William H. Donahue

Includes translation, notes, commentary, glossary, and bibliography.

1. Kepler, Johannes, 1571–1630, Astronomia Nova. 2. Astronomy–early works to 1800. 3. History of Science. 4. Humanities. 5. Science and Technology Studies.

I. Donahue, William H., 1943–. II. Title

ISBN-13: 978-1-888009-28-6 (softcover binding)

Library of Congress Control Number: 2004113309

Contents

* Chapter descriptions by the editor. Kepler's original titles are given in the text.

Editor's Preface

Kepler's *Astronomia Nova* is a single concerted argument encompassing over six hundred pages, striving to convince readers that everything they knew about astronomy is wrong. The geometrical perfection of circular orbs must be abandoned, and planetary theories must be based on a new science of physical forces. It was a hard sell. Few readers were willing to follow Kepler, but those who did found his theories, and the tables calculated from them, far superior to anything that had appeared up to that time.

In forging his "new astronomy," Kepler's deepest, most radical belief was that astronomy's task is not just to make predictions but to find the truth. As a result, he constantly raises simple questions that lead to profound inquiry. How do planets know how to move? If they are moved by forces, how do these forces work? How can simple powers be made to vary while remaining simple? How do mathematical models and analogies relate to physical and metaphysical (and even theological) reality? There are substantial sections of *Astronomia Nova* that set technical mathematical matters aside and explore these questions in a philosophical way. A selection of such chapters can thus let nonspecialist readers actually see this astronomical and physical revolution taking place before their eyes.

The aim of the present module is to present just such a selection. It is no exaggeration to say that without the thoughts revealed here, our world would have turned out to look very different to us. We will at the end arrive at Kepler's presentation of two of the three principles that we now call Kepler's Laws, but in getting there we will have a taste of the richness, cogency, and speculative brilliance of Kepler's mind.

How to read this book

The selections in this book were chosen to represent Kepler's thinking about the causes of planetary motion and, to some extent, about how to represent these causes quantitatively. How can complex phenomena be the result of simple and unchanging causes and principles? What do mathematical models and analogies really tell us? It is not necessary to read the entire book to gain some knowledge and appreciation of these themes. Here are a few suggested sequences.

Many readers will want to concentrate on Kepler's Introduction. This was the most widely disseminated of Kepler's writings in the seventeenth century and was very influential in its treatment of the interpretation of Scripture and its relation to natural science. Here

Kepler presents the physical as well as theological arguments in favor of the Copernican planetary arrangement. Kepler's Introduction can provide the basis for one or two good class discussions. Although in the part that summarizes the argument of *Astronomia Nova* there are some references to technical astronomical matters, Kepler's main points can be understood without attending to the details. Readers wishing to bypass points of technical planetary theory can either skim over them when they come up, or can begin the reading on page 8.

Those who have time and inclination to dig a little deeper might go on to read Chapters 33 and 34, in which Kepler carefully lays out his basic ideas about the causes of planetary motion. An additional class session could well be devoted to them. The text, though difficult in places, contains no mathematics, and requires only a willingness to read carefully. These chapters could well be rounded out by finishing with Chapter 57, which returns to the idea of the possible mechanism of planetary motion. It contains a few geometrical diagrams, but very little in the way of mathematics (and what there is may be omitted without loss of the sense of Kepler's argument).

The deepest level on which this module can be read involves looking into some of the reasons behind Kepler's thoughts about the causes of the motions. This sequence introduces some of the specifics of the planetary theories of Kepler's time, and shows how Kepler used simple geometry to provide hints about the real physical processes that are approximately represented by the geometrical diagrams used by Ptolemy and Copernicus. A reading on this level includes all the selections in the module, and has been completed by liberal arts students in five or six class sessions. To follow every step requires some elementary geometry and trigonometry. However, where mathematical arguments appear, it is always possible to skip to the end of the demonstration, where the conclusion is set out in nontechnical language. Editorial notes are provided to guide readers who do not wish to study the mathematical details.

If help is needed in following Kepler's references to planetary theories where they occur, Appendix A, which summarizes their main features, should be consulted. This appendix, together with the Glossary, provides a grounding in the technical terms required for understanding Kepler's text. Several other helpful books are listed in the Bibliographical Note.

William H. Donahue
January 2005

About Johannes Kepler

Kepler was born on December 27, 1571, to a locally prominent, but by no means wealthy, family in Weil der Stadt, on the edge of the Black Forest in Germany. His unusual abilities were recognized early, and he was well educated at local schools and at the University of Tübingen (with a full scholarship). Although he had wanted to be a Lutheran minister, he somewhat reluctantly took up an assignment in Graz, capital of Styria in Austria, teaching mathematics, in 1594.

His discovery (in the middle of a lecture) that the distances of the planets from the sun closely matched the spacing of a nest of Euclid's five regular solids was published in his first book, *Mysterium Cosmographicum,* in 1596. This book brought him to the attention of the Danish astronomer Tycho Brahe, who had meanwhile relocated to Prague. Accepting Brahe's invitation to come work with him on planetary theory, and driven by the expulsion of Lutherans from Graz, Kepler moved to Prague. Here he had access to some of Brahe's superb (and jealously guarded) observations, and began the work that would lead to the writing of *Astronomia Nova.* (For Kepler's description of these events, see Chapter 7, below.) During this time, he also published his *Optics* (1604), to which all subsequent study of light and vision traces its origins.

Kepler's later work, done largely under princely patronage, consolidated what he had begun in *Mysterium Cosmographicum* and *Astronomia Nova.* He also made some remarkable excursions into other areas—for example, his study of the volume of wine barrels (*Stereometria,* 1615), which anticipates certain methods of calculus. The book that he thought his best (though we might not agree) was *Harmonice Mundi* (1618), in which he tried to show that the universe is constructed on the basis of musical harmonies. His belief in universal harmony contrasts strikingly with his life, which was marked by religious strife, frequent illness and death of family members, protracted war, and a lengthy trial of his mother for witchcraft.

In 1627, he published the *Tabulae Rudolphinae,* the work that did the most to establish the superiority of his theories over those of contemporary astronomical theorists. Kepler died in 1630 at age 58, away from home in Regensburg, while attempting to meet with a patron and to collect some of the debt owed him from the treasury at Linz.

Acknowledgments

The present book originated for use in the mathematics tutorials at St. John's College in Santa Fe. For many years, the lack of a suitable selection from Kepler's writings had been acutely felt by both students and faculty. On several occasions the editor of the present module provided annotated selections from his complete translation of *Astronomia Nova* to fill this need, and the selections were used in tutorials with generally good results. Those selections tended to be technical in nature, continuing the astronomical sequence that began with Ptolemy and continued with Copernicus. There was, however, another possible approach, namely, to select readings which, while drawing on earlier astronomical readings, were primarily forward looking, showing Kepler's attempts at creating a physically based astronomy.

This focus on physical astronomy was adopted for the sophomore mathematics classes in 2003 and 2004. As a result of this collective experience with over two hundred students, the selection was revised and many more notes and diagrams were provided. Green Lion Press is deeply grateful to mathematics archons John S. Steadman and Howard J. Fisher for their assistance in selecting and refining these readings and for the insightful comments they offered. Additionally, we thank all the other teachers and students involved for their help in making this book better focused and easier to use.

The more technical set of readings and notes used in previous years was developed with the support of a grant from the National Endowment for the Humanities. Although the present selection was not developed under this grant, it evolved in part as a consequence of work done under NEH funding. This support is gratefully acknowledged.

Green Lion Press's chief editor Dana Densmore and associate editor Howard J. Fisher both read numerous drafts of this book, and they have contributed immeasurably to the final version. The editor owes more to them than he can say.

Kepler's Introduction to *Astronomia Nova*

Kepler's introduction to the *Astronomia Nova,* especially the part that discusses the relation between scripture and astronomy, was the most widely disseminated of his works during the seventeenth century. It was reprinted as an addendum to the Latin translation of Galileo's *Dialogue on the Two Chief World Systems* (Strasburg 1635 and other editions), and was the only writing of Kepler's to appear in English prior to the nineteenth century. It consists of a summary of *Astronomia Nova,* with a rather long digression on physics, gravity, tides, Scripture, and Copernicanism. Much of this digression was originally intended as an introduction to Kepler's first published work, *Mysterium Cosmographicum* (Tübingen 1596). However, the faculty of the University of Tübingen, whose approval was required, found Kepler's scriptural interpretation objectionable and demanded that the offending matter be removed. By the time of writing of *Astronomia Nova,* however, Kepler's status as Imperial Mathematician gave him more freedom to publish as he saw fit. He therefore revived his scriptural argument and inserted it into the present introduction.

Kepler's account of gravity may appear surprisingly modern, an anticipation of Newton's universal gravitation. We should beware of jumping to hasty conclusions, however. In Kepler's universe, the earth and the moon occupy the same orbit around the sun, and are thus kindred bodies, and attract one another. There is not the least notion that gravity extends to any bodies other than the earth and the moon. In Kepler's universe, the planets are not attracted to each other or to the sun. Instead, each planet is inclined by nature to be at a particular distance from the sun, which is determined by the magnitudes of concentrically arranged regular geometrical solids (for which see the notes to Chapter 7, below).

The following is Kepler's text. The section headings and sub-headings were included in the Latin edition as marginal notes. Since they clearly function as headings, they have been presented as such in the present edition. Some of the footnotes are Kepler's: these have been set off from editorial notes by an asterisk and by use of the same font as used for Kepler's text. Editorial notes are set off by a rule and are in the same font as is this introductory note.

On the difficulty of reading and writing astronomical books.

It is extremely hard these days to write mathematical books, especially astronomical ones. For unless one maintains the truly rigorous sequence of proposition, construction, demonstration, and conclusion,[1] the book will not be mathematical; but maintaining that sequence makes the reading most tiresome, especially in Latin, which lacks the articles and that gracefulness possessed by Greek when it is expressed in written symbols. Moreover, there are very few suitably prepared readers these days: the rest generally reject such works. How many mathematicians are there who put up with the trouble of working through the *Conics* of Apollonius of Perga? And yet that subject matter is the sort of thing that can be expressed much more easily in diagrams and lines than can astronomy.

I myself, who am known as a mathematician, find my mental forces wearying when, upon rereading my own work, I recall from the diagrams the sense of the proofs, which I myself had originally introduced from my own mind into the diagrams and the text. But then when I remedy the obscurity of the subject matter by inserting explanations, it seems to me that I commit the opposite fault, of waxing verbose in a mathematical context.

Furthermore, prolixity of phrases has its own obscurity, no less than terse brevity. The latter evades the mind's eye while the former distracts it; the one lacks light while the other overwhelms with superfluous glitter; the latter does not arouse the sight while the former quite dazzles it.

These considerations led me to the idea of including a kind of elucidating introduction to this work, to assist the reader's comprehension as much as possible.

I conceived this introduction as having two parts. In the first I present a synoptic table of all the chapters in the book. I think this is going to be useful, because the subject matter is unfamiliar to most people, and the various terms and various procedures used here are very much alike, and are closely related, both in general and in specific details. So when all the terms and all the procedures are juxtaposed and presented in a single display, they will be mutually explanatory. For example, I discuss the natural causes that led the ancients, though ignorant of them, to suppose an equant circle or equalizing point.[2]

1. Proposition, construction, demonstration, and conclusion are the parts of a Euclidean theorem, as formalized by Proclus.

2. For terms such as "equant" and "inequality," see the Glossary. Kepler's treatment of the equant is one of the major themes of the present selection. As we shall see in Chapter 32, Kepler thought that the equant, which varied

However, I do this in two places, namely, in Parts Three and Four. A reader who encounters this subject in Part Three might think I am dealing here with the first inequality, which is a property of the motions of each of the planets individually. And indeed, this is the case in Part Four. However, in the third part, as the summary indicates, I am discussing that equant which, under the name of the second inequality, varies the motion of all the planets in common, and primarily governs the theory of the sun. Thus the synoptic table will serve to make this distinction clear.

The introduction to this work is aimed at those who study the physical sciences.

Nevertheless, the synopsis will not be of equal assistance to all. There will be those to whom this table (which I present as a thread leading through the labyrinth of the work) will appear more tangled than the Gordian Knot. For their sake, therefore, there are many points that should be brought together here at the beginning which are presented bit by bit throughout the work, and are therefore not so easy to attend to in passing. Furthermore, I shall reveal, especially for the sake of those professors of the physical sciences who are irate with me, as well as with Copernicus and even with the remotest antiquity, on account of our having shaken the foundations of the sciences with the motion of the earth—I shall, I say, reveal faithfully the intent of the principal chapters which deal with this subject, and shall propose for inspection all the principles of the proofs upon which my conclusions, so repugnant to them, are based.

For when they see that this is done faithfully, they will then have the free choice either of reading through and understanding the proofs themselves with much exertion, or of trusting me, a professional mathematician, concerning the sound and geometrical method presented. Meanwhile, they, for their part, will turn to the principles of the proofs thus gathered for their inspection, and will examine

---

(footnote continued)

a planet's speed on its eccentric circle, was a geometrical stand-in for a dynamic principle. For this principle to be universal, it would have to hold true for the earth's orbit (which creates the second inequality) as well as for the orbits of the other planets (where it affects the first inequality). Therefore, the causes of the nonuniform motions are discussed in two different parts of the book: first, in Part 3, for the earth's orbit (see Ch. 32 and 40), and second, in Part 4, for Mars (see Ch. 44).

them thoroughly, knowing that unless they are refuted the proof erected upon them will not topple. I shall also do the same where, as is customary in the physical sciences, I mingle the probable with the necessary and draw a plausible conclusion from the mixture. For since I have mingled celestial physics with astronomy in this work, no one should be surprised at a certain amount of conjecture. This is the nature of physics, of medicine, and of all the sciences which make use of other axioms besides the most certain evidence of the eyes.

On the schools of thought in astronomy.

The reader should be aware that there are two schools of thought among astronomers, one distinguished by its chief, Ptolemy, and by the assent of the large majority of the ancients, and the other attributed to more recent proponents, although it is the most ancient. The former treats the individual planets separately and assigns causes to the motions of each in its own orb, while the latter relates the planets to one another, and deduces from a single common cause those characteristics that are found to be common to their motions. The latter school is again subdivided. Copernicus, with Aristarchus of remotest antiquity, ascribes to the translational motion of our home the earth the cause of the planets' appearing stationary and retrograde. Tycho Brahe, on the other hand, ascribes this cause to the sun, in whose vicinity he says the eccentric circles of all five planets are connected as if by a kind of knot (not physical, of course, but only quantitative). Further, he says that this knot, as it were, revolves about the motionless earth, along with the solar body.

For each of these three opinions concerning the world[3] there are several other peculiarities which themselves also serve to distinguish these schools, but these peculiarities can each be easily altered and amended in such a way that, so far as astronomy, or the celestial appearances, are concerned, the three opinions are for practical purposes equivalent to a hair's breadth, and produce the same results.

The twofold aim of the work.

My aim in the present work is chiefly to reform astronomical theory (especially of the motion of Mars) in all three forms of hypotheses, so that what we compute from the tables may correspond to the celestial

---

3. Latin, *Mundus*. This comprises the entire corporeal universe, including the fixed stars.

phenomena. Hitherto, it has not been possible to do this with sufficient certainty. In fact, in August of 1608, Mars was a little less than four degrees beyond the position given by calculation from the Prutenic tables. In August and September of 1593 this error was a little less than five degrees, while in my new calculation the error is entirely suppressed.

On the physical causes of the motions.

Meanwhile, although I place this goal first and pursue it cheerfully, I also make an excursion into Aristotle's *Metaphysics,* or rather, I inquire into celestial physics and the natural causes of the motions. The eventual result of this consideration is the formulation of very clear arguments showing that only Copernicus's opinion concerning the world (with a few small changes) is true, that the other two are false, and so on.

Indeed, all things are so interconnected, involved, and intertwined with one another that after trying many different approaches to the reform of astronomical calculations, some well trodden by the ancients and others constructed in emulation of them and by their example, none other could succeed than the one founded upon the motions' physical causes themselves, which I establish in this work.

The first step: The planes of the eccentrics intersect in the sun.

Now my first step in investigating the physical causes of the motions was to demonstrate that [the planes of] all the eccentrics intersect in no other place than the very center of the solar body (not some nearby point), contrary to what Copernicus and Brahe thought. If this correction of mine is carried over into the Ptolemaic theory, Ptolemy will have to investigate not the motion of the center of the epicycle, about which the epicycle proceeds uniformly,[4] but the motion of some point whose distance from that center, in proportion to the diameter,

---

4. In the theory of Mars, Ptolemy's epicycle actually represents the sun's motion, which is carried out uniformly about the epicycle's center. With Kepler's correction, the line of intersection of the planes of the epicycle and the eccentric would in general not pass through the center of the epicycle, but through the other nearby point described by Kepler. Thus when he says "about which the epicycle proceeds uniformly" (below), he means that the planet is proceeding uniformly on the epicycle about that point.

For more on how solar theory is implicated in planetary models, see Appendix A.

is the same as the distance of the center of the solar orb from the earth for Ptolemy, which point is also on the same line, or one parallel to it.

Here the Braheans could have raised the objection against me that I am a rash innovator, for they, while holding to the opinion received from the ancients and placing the intersection of the [planes of the] eccentrics not in the sun but near the sun, nevertheless construct on this basis a calculation that corresponds to the heavens. And in translating the Brahean numbers into the Ptolemaic form, Ptolemy could have said to me that as long as he upheld and expressed the phenomena, he would not consider any eccentric other than the one described by the center of the epicycle, about which the epicycle proceeds uniformly. Therefore I have to look again and again at what I am doing, so as to avoid setting up a new method which would not do what was already done by the old method.

So to counter this objection, I have demonstrated in the first part of the work [Chapters 1–6] that exactly the same things can result or be presented by this new method as are presented by their ancient method.

In the second part of the work [Chapters 7–21] I take up the main subject, and describe the positions of Mars at apparent opposition to the sun, not worse, but indeed much better, with my method than they expressed the positions of Mars at mean opposition to the sun with the old method.

Meanwhile, throughout the entire second part (as far as concerns geometrical demonstrations from the observations) I leave in suspense the question of whose procedure is better, theirs or mine, seeing that we both match a great many observations (this is, indeed, a basic requirement for our theorizing). However, my method is in agreement with physical causes, and their old one is in disagreement, as I have partly shown in the first part, especially Chapter 6.

But finally in the fourth part of the work [Chapters 41–60], in Chapter 52, I consider certain other observations, no less trustworthy than the previous ones were, which their old method could not match, but which mine matches most beautifully. I thereby demonstrate most soundly that Mars's eccentric is so situated that the center of the solar body lies upon its line of apsides, and not any nearby point, and hence, that all the [planes of the] eccentrics intersect in the sun itself.

This should, however, hold not just for the longitude, but for the latitude as well. Therefore, in the fifth part [Chapters 61–70] I have demonstrated the same from the observed latitudes, in Chapter 67.

The second step: there is an equant in the theory of the sun

This could not have been demonstrated earlier in the work, because one of the constituents of these astronomical demonstrations is an exact knowledge of the causes of the second inequality in the planets' motion, for which some other new thing had likewise to be discovered in the third part, unknown to our predecessors, and so on.

For I have demonstrated in the third part [Chapters 22–40] that whether the old method, which depends upon the sun's mean motion, is valid, or my new one, which uses the apparent motion, nevertheless, in either case there is something from the causes of the first inequality that is mixed in with the second, which pertains to all planets in common. Thus for Ptolemy I have demonstrated that his epicycles do not have as centers those points about which their motion is uniform. Similarly for Copernicus I have demonstrated that the circle in which the earth is moved around the sun does not have as its center that point about which its motion is regular and uniform. Similarly for Tycho Brahe I have demonstrated that the circle on which the common point or knot of the eccentrics mentioned above moves does not have as its center that point about which its motion is regular and uniform. For if I concede to Brahe that the common point of the eccentrics may be different from the center of the sun, he must grant that the circuit of that common point, which in magnitude and period exactly equals the orbit of the sun, is eccentric and tends towards Capricorn, while the sun's eccentric circuit tends towards Cancer. The same thing befalls Ptolemy's epicycles.

However, if I place the common point or knot of the eccentrics in the center of the solar body, then the common circuit of both the knot and the sun is indeed eccentric with respect to the earth, and tends towards Cancer, but by only half the eccentricity shown by the point about which the sun's motion is regular and uniform.

And in Copernicus, the earth's eccentric still tends towards Capricorn, but by only half the eccentricity of the point about which the earth's motion is uniform, also in the direction of Capricorn.

Likewise, in Ptolemy, on each of the diameters of the epicycles that run from Capricorn to Cancer, there are three points, the outer two of which are at the same distance from the middle ones; and their distances from one another have the same ratio to the diameters as the whole eccentricity of the sun has to the diameter of its circuit. And of these three points, the middle ones are the centers of their epicycles, those that lie toward Cancer are the points about which the motions on the epicycles are uniform, and finally those that lie toward Capricorn are the ones whose eccentrics (described by them) we would be tracing out *if instead of the sun's mean motion we follow the apparent motion,* just as if those were the points at which the epicycles were attached to the eccentric. The result of this is that each planetary epicycle contains the theory of the sun in its entirety, with all the properties of its motions and circles.

The earth is moved and the sun stands still. Physico-astronomical arguments.

With these things thus demonstrated by a reliable method, the previous step towards the physical causes is now confirmed, and a new step is taken towards them, most clearly in the theories of Copernicus and Brahe, and more obscurely but at least plausibly in the Ptolemaic theory.

For whether it is the earth or the sun that is moved, it has certainly been demonstrated that the body that is moved is moved in a nonuniform manner, that is, slowly when it is farther from the body at rest, and more swiftly when it has approached this body.

Thus the physical difference is now immediately apparent—by way of conjecture, it is true, but yielding nothing in certainty to conjectures of doctors on physiology or to any other natural science.

First, Ptolemy is certainly condemned. For who would believe that there are as many theories of the sun (so closely resembling one another that they are in fact equal) as there are planets, when he sees that for Brahe a single solar theory suffices for the same task, and it is the most widely accepted axiom in the natural sciences that Nature makes use of the fewest possible means?

That Copernicus is better able than Brahe* to deal with celestial physics is proven in many ways.

1. First, although Brahe did indeed take up those five solar theories from the theories of the planets, bringing them down to the centers of the eccentrics, hiding them there, and conflating them into one, he nevertheless left in the world the effects produced by those theories. For Brahe no less than for Ptolemy, besides that motion which is proper to it, each planet is still actually moved with the sun's motion, the two being mixed into one, the result being a spiral. That it results from this that there are no solid orbs, Brahe has demonstrated most

* Of whom, in all fairness, most honest and grateful mention is made, and recognition given, since I build this entire structure from the bottom up upon his work, all the materials being borrowed from him. [Kepler's note]

firmly. Copernicus, on the other hand, entirely removed this extrinsic motion from the five planets, assigning its cause to a deception arising from the circumstances of observation. Thus the motions are still multiplied to no purpose by Brahe, as they were before by Ptolemy.

2. Second, if there are no orbs, the conditions under which the intelligences and moving souls must operate are made very difficult, since they have to attend to so many things to introduce to the planet two intermingled motions. They would at least have to attend at one and the same time to the principles, centers, and periods of the two motions. But if the earth is moved, I show that most of this can be done with physical rather than animate faculties,[5] namely, magnetic ones. But these are more general points. There follow others arising specifically from demonstrations, upon which we now begin.

3. For if the earth is moved, it has been demonstrated that the increases and decreases of its velocity are governed by its approaching towards and receding from the sun. And in fact the same happens with the rest of the planets: they are urged on or held back according to the approach toward or recession from the sun. So far, the demonstration is geometrical.

And now, from this very reliable demonstration, the conclusion is drawn, using a physical conjecture, that the source of the five planets' motion is in the sun itself. It is therefore very likely that the source of the earth's motion is in the same place as the source of the other five planets' motion, namely, in the sun as well. It is therefore likely that the earth is moved, since a likely cause of its motion is apparent.

4. That, on the other hand, the sun remains in place in the center of the world, is most probably shown by (among other things) its being the source of motion for at least five planets. For whether you follow Copernicus or Brahe, the source of motion for five of the planets is in the sun, and in Copernicus, for a sixth as well, namely, the earth. And it is more likely that the source of all motion should remain in place rather than move.

5. But if we follow Brahe's theory and say that the sun moves, this first conclusion still remains valid, that the sun moves slowly when it is more distant from the earth and swiftly when it approaches, and this not only in appearance, but in fact. For this is the effect of the circle of the equant, which, by an inescapable demonstration, I have introduced into the theory of the sun.

5. Latin, *facultas animalis.* This is the faculty of the soul that controls local motion in animals.

Upon this most valid conclusion, making use of the physical conjecture introduced above, might be based the following theorem of natural philosophy: the sun, and with it the whole huge burden (to speak coarsely) of the five eccentrics, is moved by the earth; or, the source of the motion of the sun and the five eccentrics attached to the sun is in the earth.

Now let us consider the bodies of the sun and the earth, and decide which is better suited to being the source of motion for the other body. Does the sun, which moves the rest of the planets, move the earth, or does the earth move the sun, which moves the rest, and which is so many times greater? Unless we are to be forced to admit the absurd conclusion that the sun is moved by the earth, we must allow the sun to be fixed and the earth to move.

6. What shall I say of the motion's periodic time of 365 days, intermediate in quantity between the periodic time of Mars of 687 days and that of Venus of 225 days? Does not the nature of things cry out with a great voice that the circuit in which these 365 days are used up[6] also occupies a place intermediate between those of Mars and Venus about the sun, and thus itself also encircles the sun, and hence, that this circuit is a circuit of the earth about the sun, and not of the sun about the earth? These points are, however, more appropriate to my *Mysterium cosmographicum*, and arguments that are not going to be repeated in this work should not be introduced here.

7. For other metaphysical arguments that favor the sun's position in the center of the world, derived from its dignity or its illumination, see my little book just mentioned, or look in Copernicus. There is also something in Aristotle's *On the Heavens*, Book II, in the passage on the Pythagoreans, who used the name "fire" to signify the sun.[7] I have touched upon a few points in the *Astronomiae pars optica* Ch. 1 p. 7, and also Ch. 6, especially p. 225.[8]

---

6. Behind this odd phrase lies Kepler's peculiar treatment of time as a dependent variable: he makes consistent use of the amounts of time "used up" in traversing equal distances, rather than considering the distances traversed in equal times (as Galileo and his successors did). It is quite likely that this different viewpoint was of importance in developing the "area law" which later became known as Kepler's Second Law. See especially the beginning of Chapter 40.

7. Aristotle, *On the Heavens*, II.13, 293a 20.

8. Kepler, *Optics*, p. 20 in the Donahue translation: "The sun is accordingly a particular body; in it is this faculty of communicating itself to all things,

8. But on the earth's being suited to a circular motion in some place other than the center of the world, you will find a metaphysical argument in Chapter 9, p. 322 of that book.[9]

Objections to the earth's motion.

I trust the reader's indulgence if I take this opportunity to present a few brief replies to a number of objections which, capturing people's minds, use the following arguments to shed darkness. For these replies are by no means irrelevant to matters that concern the physical causes of the planets' motion, which I discuss chiefly in parts three and four of the present work [Chapters 22–60].

I. On the motion of heavy bodies.

Many are prevented by the motion of heavy bodies from believing that the earth is moved by an animate motion, or better, by a magnetic one. They should ponder the following propositions.

The theory of gravity is in error.

A mathematical point, whether or not it is the center of the world, can neither effect the motion of heavy bodies nor act as a object towards which they tend. Let the physicists prove that this force is in

---

(Footnote continued)

which we call light, to which, on this account at least, is due the middle place in the whole world, and the center, so that it might perpetually pour itself forth equably into the whole orb."
Optics, Donahue trans. p. 241: "You will see in the sun a palpable image of the world; in the world, that of God the Creator."

9. Kepler, *Optics*, Donahue trans. pp. 331–2: "And so it was evidently not fitting that the human being, destined to be the inhabitant and watchman of this world, should reside in its middle, as if in a closed cubicle, under which circumstance he would never have made his way through to the contemplation of heavenly bodies that are so remote; but rather, by the annual translatory motion of the earth, his domicile, he circumambulates and strolls around in this most ample building, so as to be able more rightly to perceive and measure the individual members of the house."

Many of Kepler's most perceptive insights originate in his consistency in viewing planetary motions in three dimensions rather than as circles in a plane. In this passage from the *Optics*, we see how, in Kepler's view, God deliberately placed us on a moving platform so that we could make good use of our spatial sense. Kepler's use of his remarkable sense of space is one of the guiding themes of the present selection.

a point which neither is a body nor is grasped otherwise than through mere relation.

It is impossible that, in moving its body, the form of a stone seek out a mathematical point (in this instance, the center of the world), without respect to the body in which this point is located. Let the physicists prove that natural things have a sympathy for that which is nothing.

It is likewise impossible for heavy bodies to tend towards the center of the world simply because they are seeking to avoid its spherical extremities. For, compared with their distance from the extremities of the world, the proportional part by which they are removed from the world's center is imperceptible and of no account. Also, what would be the cause of such antipathy? With how much force and wisdom would heavy bodies have to be endowed in order to be able to flee so precisely an enemy surrounding them on all sides? Or what ingenuity would the extremities of the world have to possess in order to pursue their enemy with such exactitude?

Nor are heavy bodies driven in towards the middle by the rapid whirling of the *primum mobile,* as objects in whirlpools are. That motion (if we suppose it to exist) does not carry all the way down to these lower regions. If it did, we would feel it, and would be caught up by it along with the very earth itself; indeed, we would be carried ahead, and the earth would follow. All these absurdities are consequences of our opponents' view, and it therefore appears that the common theory of gravity is in error.

True theory of gravity

The true theory of gravity rests upon the following axioms.[10]

Every corporeal[11] substance, to the extent that it is corporeal, has

---

10. As Max Caspar notes in his edition of the *Astronomia Nova,* the theory presented in these terse statements constitutes a complete rejection of the Aristotelian view of gravity and plays an important role in Kepler's physical thought. In later works Kepler refers back to them, especially in Book I Part 4 of the *Epitome of Copernican Astronomy,* where he develops them further.

11. Latin, *corporea.* There is a close relation in Kepler's thought here between *corpus* (body) and *corporea* that is not made entirely clear in this translation. Other possible renderings would be "physical" or "bodily." The former would not adequately represent Kepler's meaning in the first axiom, while the latter is sufficiently at odds with correct usage to lead the translator to reject it, though with some regret.

been so made as to be suited to rest in every place in which it is put by itself, outside the sphere of influence of a kindred[12] body.

Gravity is a mutual corporeal disposition among kindred bodies to unite or join together; thus, the earth attracts a stone much more than the stone seeks the earth. (The magnetic faculty is another example of this sort.)

Heavy bodies (most of all if we establish the earth in the center of the world) are not drawn towards the center of the world because it is the center of the world, but because it is the center of a kindred spherical body, namely, the earth. Consequently, wherever the earth be established, or whithersoever it be carried by its animate faculty, heavy bodies are drawn towards it.

If the earth were not round, heavy bodies would not everywhere be drawn in straight lines towards the middle point of the earth, but would be drawn towards different points from different sides.

If two stones were set near one another in some place in the world outside the sphere of influence of a third kindred body, these stones, like two magnetic bodies, would come together in an intermediate place, each approaching the other by an interval proportional to the bulk [*moles*] of the other.

If the moon and the earth were not each held back in its own circuit by an animate force or something else equivalent to it, the earth would ascend towards the moon by one fifty-fourth part of the interval, and the moon would descend towards the earth about fifty-three parts of the interval, and there they would be joined together; provided, that is, that the substance of each is of the same density.[13]

If the earth should cease to attract its waters to itself, all the sea water would be lifted up, and would flow onto the body of the moon.

Reason for the ebb and flow of the sea.

The sphere of influence of the attractive power in the moon is

---

12. Latin, *cognata,* "of the same origin."

13. The apparent angular size of the moon is about half a degree, so the apparent size of its radius is about 1/4 degree. The moon's distance from earth is about 60 times the earth's radius (these numbers were well known to the ancient Greeks). By elementary trigonometry, 60 times the sine of 1/4 degree will give the ratio of the radii of earth and moon, which comes out to a little less than 1:4. Cubing this gives the ratio of volumes, to which, Kepler supposes, the attractive power is proportional.

extended all the way to the earth, and in the torrid zone calls the waters forth, particularly when it comes to be overhead in one or another of its passages. This is imperceptible in enclosed seas, but noticeable where the beds of the ocean are widest and there is much free space for the waters' reciprocation. It thus happens that the shores of the temperate latitudes are laid bare, and to some extent even in the torrid regions the neighboring oceans diminish the size of the bays. And thus when the waters rise in the wider ocean beds, the moon being present, it can happen that in the narrower bays, if they are not too closely surrounded, the water might even seem to be fleeing the moon, though in fact they are subsiding because a quantity of water is being carried off elsewhere.

But the moon passes the zenith swiftly, and the waters are unable to follow so swiftly. Therefore, a current arises in the ocean of the torrid zone, which, when it strikes upon the far shores, is thereby deflected. But when the moon departs, this congress of the waters, or army on the march towards the torrid zone, now abandoned by the traction that had called it forth, is dissolved. But since it has acquired impetus, it flows back (as in a water vessel) and assaults its own shore, inundating it. In the moon's absence, this impetus gives rise to another impetus until the moon returns and the impetus is restrained, moderated, and carried along with the moon's motion. So all shores that are equally accessible are flooded at the same time, while those more remote are flooded later, some in different ways because of their various degrees of accessibility to the ocean.

Effects of the sea's ebb and flow.

I will point out in passing that the sand dunes of the Syrtes[14] are heaped up in this way; that thus are created or destroyed countless islands in bays full of eddies (such as the Gulf of Mexico); that it seems that the soft, fertile, and friable earth of the [East] Indies was thus at length broached and penetrated by this current, this perpetual inundation, with help from a certain all-pervading motion of the earth. For it is said that India was once continuous from the Golden

14. Shoals on the coast of North Africa. Kepler is probably referring to Syrtis Minor, the Gulf of Gabès, in Tunisia. This is one of the few parts of the Mediterranean that has appreciable tides, and extensive sandy shoals are exposed at low tide.

Chersonese[15] towards the east and south, but now the ocean, which was once farther back between China and America, has flowed in, and the shores of the Moluccas and of other neighboring islands, which are now raised on high because of the subsidence of the surface of the sea, bear witness to this event.

Taprobane,[16] too, seems to have been submerged through this cause (as is consistent with the account of the Calcuttans that several localities there were once submerged), when the China Sea burst in through breaches into the Indian Ocean, with the result that nowadays nothing of Taprobane remains but the peaks of the mountains, which take the form of the innumerable islands known as the Maldives. For it is easy to prove, from the geographers and Diodorus Siculus, that that was once the site of Taprobane, namely, to the south opposite the mouths of the Indus and the promontory of Corium.[17] Moreover, in ecclesiastical history one individual is said to have been bishop of Arabia and Taprobane together, and so the latter must surely have been nearby and not five hundred German miles to the east (indeed, more than a thousand, following the roundabout routes used in those days). The island of Sumatra, nowadays considered to be Taprobane, I think was once the Golden Chersonese, joined to the Indian isthmus at the city of Malacca. For Chersonesus,[18] which nowadays we believe to be the Golden, seems to have no more right than Italy to the name "Chersonese."

Although these things are appropriate to a different topic, I wanted to present them all in one context in order to make more credible the ocean tide and through it the moon's attractive power.

For it follows that if the moon's power of attraction extends to the earth, the earth's power of attraction will be much more likely to extend to the moon and far beyond, and accordingly, that nothing that consists to any extent whatever of terrestrial material, carried up on high, ever escapes the grasp of this mighty power of attraction.

---

15. "Chersonese" is a Greek-derived word equivalent to the Latin-derived "peninsula." The Golden Chersonese is usually identified with the Malay Peninsula.

16. This name usually refers to the island now known as Sri Lanka, though Kepler has his doubts.

17. A search in Zedler's massive *Grosses Universal-Lexicon* suggests that Corium may be in the province of Iran now known as Kerman.

18. The Thracian Chersonese (Gallipoli).

True theory of levity.

Nothing that consists of corporeal material is absolutely light. It is only comparatively lighter, because it is less dense, either by its own nature or through an influx of heat. By "less dense" I do not just mean that which is porous and divided into many cavities, but in general that which, while occupying a place of the same magnitude as that occupied by some heavier body, contains a lesser quantity of corporeal material.

The motion of light things also follows from their definition. For it should not be thought that they flee all the way to the surface of the world when they are carried upwards, or that they are not attracted by the earth. Rather, they are less attracted than heavy bodies, and are thus displaced by heavy bodies, whereupon they come to rest and are kept in their place by the earth.

To the objection that objects projected vertically fall back to their places.

But even if the earth's power of attraction is extended very far upwards, as was said, nevertheless, if a stone were at a distance that was perceptible in relation to the earth's diameter, it is true that, the earth being moved, such a stone would not simply follow, but its forces of resistance would mingle with the earth's forces of attraction, and it would thus detach itself somewhat from the earth's grasp. In just the same way, violent[19] motion detaches projectiles somewhat from the earth's grasp, so that they either run on ahead if they are shot eastwards, or are left behind if shot westwards, thus leaving the place from which they are shot, under the compulsion of force. Nor can the earth's revolving effect impede this violent motion all at once, as long as the violent motion is at its full strength.

But no projectile is separated from the surface of the earth by even a hundred thousandth part of the earth's diameter, and not even the clouds themselves, or smoke, which partake of earthy matter to

19. A technical Aristotelian term, for which there is no satisfactory modern equivalent. Aristotle categorized all motions as being either "natural" or "violent," depending upon whether they are carried out in accordance with some inner principle or are caused by something external. Here, the "natural" motion is the coming together of all terrestrial bodies, while the "violent" motion is the separation of those bodies. Kepler is arguing that when a body is separated from kindred bodies, there is some faculty that brings into action a force tending to bring the bodies back together.

the very least extent, achieve an altitude of a thousandth part of the semidiameter.[20] Therefore, none of the clouds, smokes, or objects shot vertically upwards can make any resistance, nor, I say, can the natural inclination to rest do anything to impede this grasp of the earth's, at least where this resistance is negligible in proportion to that grasp. Consequently, anything shot vertically upwards falls back to its place, the motion of the earth notwithstanding. For the earth cannot be pulled out from under it, since the earth carries with it anything sailing through the air, linked to it by the magnetic force no less firmly than if those bodies were actually in contact with it.

When these propositions have been grasped by the understanding and pondered carefully, not only do the absurdity and falsely conceived physical impossibility of the earth's motion vanish, but it also becomes clear how to reply to the physical objections, however they are framed.

The opinion of Copernicus.

Copernicus preferred to think that the earth and all terrestrial bodies (even those cast away from the earth) are informed by one and the same motive soul, which, while rotating its body the earth, also rotates those particles cast away from it. He thus held it to be this soul, spread throughout the particles, that acquires force through violent motions, while I hold that it is a corporeal faculty (which we call gravity, or the magnetic faculty), that acquires the force in the same way, namely, through violent motions.

Nevertheless, this corporeal faculty is sufficient for anything removed from the earth: the animate faculty is superfluous.

II. To objections concerning the swiftness of the earth's motion.

Although many people fear the worst for themselves and for all earth's creatures on account of the extreme rapidity of this motion, they have no cause for alarm. On this point see my book, *On the New Star,* Chapters 15 and 16, pp. 82 and 84.

20. In the *Optics*, Ch. 4 Prop. 11, Donahue trans. pp. 141–2, Kepler argued on the basis of atmospheric refraction that the altitude of the air cannot be greater than half a German mile (about 3.7 km), and therefore concluded that the highest mountain peaks must be above the atmosphere. His chief error was to suppose that the air maintains its density with increasing altitude, and has a well-defined surface, as water does.

III. To objections concerning the immensity of the heavens.

In the same place, you will find the full-sail voyage along the world's immense orbit, which, in objection to Copernicus, is usually held to be unnatural. There it is demonstrated to be well-proportioned, and that, on the contrary, the speed of the heavens would become ill-proportioned and unnatural were the earth ordered to remain quite motionless in its place.

IV. To objections concerning the dissent of holy scripture, and its authority.[21]

There are, however, many more people who are moved by piety to withhold assent from Copernicus, fearing that falsehood might be charged against the Holy Spirit speaking in the scriptures if we say that the earth is moved and the sun stands still.

But let them consider that since we acquire most of our information, both in quality and quantity, through the sense of sight, it is impossible for us to abstract our speech from this ocular sense. Thus, many times each day we speak in accordance with the sense of sight, although we are quite certain that the truth of the matter is otherwise. This verse of Virgil furnishes an example:

> We are carried from the port, and the land and cities recede.[22]

Thus, when we emerge from the narrow part of some valley, we say that a great plain is opening itself out before us.

Thus Christ said to Peter, "Lead forth on high,"[23] as if the sea were higher than the shores. It does seem so to the eyes, but optics shows the cause of this fallacy. Christ was only making use of the common idiom, which nonetheless arose from this visual deception.

---

21. The following arguments on the interpretation of scripture were to become the most widely read of Kepler's writings. They were often reprinted from the seventeenth century on, and translated into modern languages. Indeed, this part of the Introduction was the only work of Kepler's to appear in English before the 1870s.

22. *Aeneid* III. 72. This line was also quoted by Copernicus, Book I Chapter 8 of *De revolutionibus*.

23. Luke 5:4. The Latin *altum* can mean either "high" or "deep." However, Kepler cannot have been unaware that the original Greek verse unambiguously has the latter meaning, and hence he must be charged with making a rather silly distortion in order to prove a point.

Thus, we call the rising and setting of the stars "ascent" and "descent," though at the same time that we say the sun ascends, others say it descends. See the *Optics* Ch. 10 p. 327.[24]

Thus, the Ptolemaic astronomers even now say that the planets are stationary when they are seen to stay near the same fixed stars for several days, even though they think the planets are then really moving downwards in a straight line, or upwards away from the earth.

Thus writers of all nations use the word "solstice" [sun-standing, in Latin] even though they in fact deny that the sun stands still.

Thus there has not yet been anyone so doggedly Copernican as to avoid saying that the sun is entering Cancer or Leo, even though he wishes to signify that the earth is entering Capricorn or Aquarius. And there are other like examples.

Now the holy scriptures, too, when treating common things (concerning which it is not their purpose to instruct humanity), speak with humans in the human manner, in order to be understood by them. They make use of what is generally acknowledged, in order to weave in other things more lofty and divine.

No wonder, then, if scripture also speaks in accordance with human perception when the truth of things is at odds with the senses, whether or not humans are aware of this. Who is unaware that the allusion in Psalm 19 [verses 5–7] is poetical? Here, under the image of the sun, are sung the spreading of the Gospel and even the sojourn of Christ the Lord in this world on our behalf, and in the singing the sun is said to emerge from the tabernacle of the horizon like a bridegroom from his marriage bed, exuberant as a strong man for the race. Which Virgil imitates thus:

> Aurora leaving Tithonus's saffron-colored bed.[25]

The psalmodist was aware that the sun does not go forth from the horizon as from a tabernacle (even though it may appear so to the eyes). On the other hand, he considered the sun to move for the precise reason that it appears so to the eyes. In either case, he expressed it so because in either case it appeared so to the eyes. He should not be judged to have spoken falsely in either case, for the perception of the eyes also has its truth, well suited to the psalmodist's more

---

24. Kepler, *Optics*, Donahue trans. pp. 337–8.

25. *Aeneid* IV. 585. In Kepler's time it was generally believed that Virgil was familiar with the Hebrew scriptures.

hidden aim, the adumbration of the Gospel and also of the Son of God. Likewise, Joshua [Joshua 10:12 ff] makes mention of the valleys against which the sun and moon moved, because when he was at the Jordan it appeared so to him. Yet each writer was in perfect control of his meaning. David (and Syracides[26] with him) was describing the magnificence of God made manifest, which he expressed so as to exhibit them to the eyes, and possibly also for the sake of a mystical sense spelled out through these visible things. Joshua meant that the sun should be held back in its place in the middle of the sky for an entire day with respect to the sense of his eyes, since for other people during the same interval of time it would remain beneath the earth.

But thoughtless persons pay attention only to the verbal contradiction, "the sun stood still" versus "the earth stood still," not considering that this contradiction can only arise in an optical and astronomical context, and does not carry over into common usage. Nor are these thoughtless ones willing to see that Joshua was simply praying that the mountains not remove the sunlight from him, which prayer he expressed in words conforming to the sense of sight, as it would be quite inappropriate to think, at that moment, of astronomy and of visual errors. For if someone had admonished him that the sun doesn't really move against the valley of Ajalon, but only appears to do so, wouldn't Joshua have exclaimed that he only asked for the day to be lengthened, however that might be done? He would therefore have replied in the same way if anyone had begun to present him with arguments for the sun's perpetual rest and the earth's motion.

Now God easily understood from Joshua's words what he meant, and responded by stopping the motion of the earth, so that the sun might appear to him to stop. For the gist of Joshua's petition comes to this, that it might appear so to him, whatever the reality might meanwhile be. Indeed, that this appearance should come about was not vain and purposeless, but quite conjoined with the desired effect.

But see Chapter 10 of the *Astronomiae pars optica*,[27] where you will find reasons why, to absolutely everyone, the sun appears to move and not the earth: it is because the sun appears small and the earth large, and also because, owing to its apparent slowness, the sun's motion is perceived, not by sight, but by reasoning alone, through

26. Yehoshua ben Sirach, author of the apocryphal Biblical text *Ecclesiasticus*, or *Sirach*.

27. Kepler, *Optics*, Ch. 10, Donahue trans. pp. 335–346.

its change of distance from the mountains over a period of time. It is therefore impossible for a previously uninformed reason to imagine anything but that the earth, along with the arch of heaven set over it, is like a great house, immobile, in which the sun, so small in stature, travels from one side to the other like a bird flying in the air.

What absolutely all men imagine, the first line of holy scripture presents. "In the beginning," says Moses, "God created the heaven and the earth," because it is these two parts that chiefly present themselves to the sense of sight. It is as though Moses were to say to man, "This whole worldly edifice that you see, light above and dark and widely spread out below, upon which you are standing and by which you are roofed over, has been created by God."

In another passage [Jeremiah 31:37], Man is asked whether he has learned how to seek out the height of heaven above, or the depths of the earth below, because to the ordinary man both appear to extend through equally infinite spaces. Nevertheless, there is no one in his right mind who, upon hearing these words, would use them to limit astronomers' diligence either in showing the contemptible smallness of the earth in comparison with the heavens, or in investigating astronomical distances. For these words do not concern measurements arrived at by reasoning. Rather, they concern real exploration, which is utterly impossible for the human body, fixed upon the land and drawing upon the free air. Read all of Chapter 38 of Job, and compare it with matters discussed in astronomy and in physics.

Suppose someone were to assert, from Psalm 24, that the earth is founded upon rivers, in order to support the novel and absurd philosophical conclusion that the earth floats upon rivers. Would it not be correct to say to him that he should regard the Holy Spirit as a divine messenger, and refrain from wantonly dragging Him into physics class? For in that passage the psalmodist intends nothing but what men already know and experience daily, namely, that the land, raised on high after the separation of the waters, has great rivers flowing through it and seas surrounding it. Not surprisingly, the same figure of speech is adopted in another passage, where the Israelites sing that they were seated upon the waters of Babylon [Psalm 137], that is, by the riverside, or on the banks of the Euphrates and Tigris.

If this be easily accepted, why can it not also be accepted that in other passages usually cited in opposition to the earth's motion we should likewise turn our eyes from physics to the aims of scripture?

A generation passes away (says Ecclesiastes [1:4]), and a generation comes, but the earth stands forever. Does it seem here as if Solomon wanted to argue with the astronomers? No; rather, he wanted to warn people of their own mutability, while the earth, home of the human race, remains always the same, the motion of the sun perpetually returns to the same place, the wind blows in a circle and returns to its starting point, rivers flow from their sources into the sea, and from the sea return to the sources, and finally, as these people perish, others are born. Life's tale is ever the same; there is nothing new under the sun.

You do not hear any physical dogma here. The message is a moral one, concerning something self-evident and seen by all eyes but seldom pondered. Solomon therefore urges us to ponder. Who is unaware that the earth is always the same? Who does not see the sun return daily to its place of rising, rivers perennially flowing towards the sea, the winds returning in regular alternation, and men succeeding one another? But who really considers that the same drama of life is always being played, only with different characters, and that not a single thing in human affairs is new? So Solomon, by mentioning what is evident to all, warns of that which almost everyone wrongly neglects.

It is said, however, that Psalm 104, in its entirety, is a physical discussion, since the whole of it is concerned with physical matters. And in it, God is said to have "founded the earth upon its stability, that it not be laid low unto the ages of ages."[28] But in fact, nothing could be farther from the psalmodist's intention than speculation about physical causes. For the whole thing is an exultation upon the greatness of God, who made all these things: the author has composed a hymn to God the creator, in which he treats the world in order, as it appears to the eyes.

If you consider carefully, you will see that it is a commentary upon the six days of creation in Genesis. For in the latter, the first three days are given to the separation of the regions: first, the region of light from the exterior darkness; second, the waters from the waters by the interposition of an extended region; and third, the land from the seas, where the earth is clothed with plants and shrubs. The last three days, on the other hand, are devoted to the filling of the regions so distinguished: the fourth, of the heavens; the fifth, of the seas and the air; and the sixth, of the land. And in this psalm there are likewise the same number of distinct parts, analogous to the works of the six days.

28. The Latin of the Vulgate, quoted by Kepler, differs markedly from the Hebrew (and hence from most English translations) here.

In the second verse, he enfolds the Creator with the vestment of light, first of created things, and the work of the first day.

The second part begins with the third verse, and concerns the waters above the heavens, the extended region of the heavens, and atmospheric phenomena that the psalmodist ascribes to the waters above the heavens, namely, clouds, winds, tornadoes, and lightning.

The third part begins with the sixth verse, and celebrates the earth as the foundation of the things being considered. The psalmodist relates everything to the earth and to the things that live on it, because, in the judgement of sight, the chief parts of the world are two: heaven and earth. He therefore considers that for so many ages now the earth has neither sunk nor cracked apart nor tumbled down, yet no one has certain knowledge of what it is founded upon.

He does not wish to teach things of which men are ignorant, but to recall to mind something they neglect, namely, God's greatness and potency in a creation of such magnitude, so solid and stable. If an astronomer teaches that the earth is carried through the heavens, he is not spurning what the psalmodist says here, nor does he contradict human experience. For it is still true that the land, the work of God the architect, has not toppled as our buildings usually do, consumed by age and rot; that it has not slumped to one side; that the dwelling places of living thing have not been set in disarray; that the mountains and coasts have stood firm, unmoved against the blast of wind and wave, as they were from the beginning. And then the psalmodist adds a beautiful sketch of the separation of the waters by the continents, and adorns his account by adding springs and the amenities that springs and crags provide for bird and beast. He also does not fail to mention the adorning of the earth's surface, included by Moses among the works of the third day, although the psalmodist derives it from its prior cause, namely, a humidification arising in the heavens, and embellishes his account by bringing to mind the benefits accruing from that adornment for the nurture and pleasure of humans and for the lairs of the beasts.

The fourth part begins with verse 20, and celebrates the work of the fourth day, the sun and the moon, but chiefly the benefit that the division of times brings to humans and other living things. It is this benefit that is his subject matter: it is clear that he is not writing as an astronomer here.

If he were, he would not fail to mention the five planets, than whose motion nothing is more admirable, nothing more beautiful, and nothing a better witness to the Creator's wisdom, for those who take note of it.

The fifth part, in verse 26, concerns the work of the fifth day, where he fills the sea with fish and ornaments it with sea voyages.

The sixth is added, though obscurely, in verse 28, and concerns the animals living on land, created on the sixth day. At the end, in conclusion, he declares the general goodness of God in sustaining all things and creating new things. So everything the psalmodist said of the world relates to living things. He tells nothing that is not generally acknowledged, because his purpose was to praise things that are known, not to seek out the unknown. It was his wish to invite men to consider the benefits accruing to them from each of these works of the six days.

Advice to astronomers.

I, too, implore my reader, when he departs from the temple and enters astronomical studies, not to forget the divine goodness conferred upon men, to the consideration of which the psalmodist chiefly invites. I hope that, with me, he will praise and celebrate the Creator's wisdom and greatness, which I unfold for him in the more perspicacious explanation of the world's form, the investigation of causes, and the detection of errors of vision. Let him not only extol the Creator's divine beneficence in His concern for the well-being of all living things, expressed in the firmness and stability of the earth, but also acknowledge His wisdom expressed in its motion, at once so well hidden and so admirable.

Advice for idiots.

But whoever is too stupid to understand astronomical science, or too weak to believe Copernicus without affecting his faith, I would advise him that, having dismissed astronomical studies and having damned whatever philosophical opinions he pleases, he mind his own business and betake himself home to scratch in his own dirt patch, abandoning this wandering about the world. He should raise his eyes (his only means of vision) to this visible heaven and with his whole heart burst forth in giving thanks and praising God the Creator. He can be sure that he worships God no less than the astronomer, to whom God has granted the more penetrating vision of the mind's eye, and an ability and desire to celebrate his God above those things he has discovered.

Commendation of the Brahean hypothesis.

At this point, a modest (though not too modest) commendation to the learned should be made on behalf of Brahe's opinion of the form of the world, since in a way it follows a middle path. On the one hand, it frees the astronomers as much as possible from the useless apparatus of so many epicycles and, with Copernicus, it includes the causes of motion, unknown to Ptolemy, giving some place to physical theory in accepting the sun as the center of the planetary system. And on the other hand, it serves the mob of literalists and eliminates the motion of the earth, so hard to believe, although many difficulties are thereby insinuated into the theories of the planets in astronomical discussions and demonstrations, and the physics of the heavens is no less disturbed.

V. To objections concerning the authority of the pious.

So much for the authority of holy scripture. As for the opinions of the pious[29] on these matters of nature, I have just one thing to say: while in theology it is authority that carries the most weight, in philosophy it is reason. Therefore, Lactantius is pious, who denied that the earth is round,[30] Augustine is pious, who, though admitting the roundness, denied the antipodes, and the Inquisition[31] nowadays is pious, which, though allowing the earth's smallness, denies its motion. To me, however, the truth is more pious still, and (with all due respect for the Doctors of the Church) I prove philosophically not only that the earth is round, not only that it is inhabited all the way around at the antipodes, not only that it is contemptibly small, but also that it is carried along among the stars.

But enough about the truth of the Copernican hypothesis. Let us return to the plan I proposed at the beginning of this introduction.

I had begun to say that in this work I treat all of astronomy by means of physical causes rather than fictitious hypotheses, and that I

---

29. Latin, *Sancti*, literally, "the Saints," or "the Holy." Context shows, however, that in modern usage "pious" or "saintly" fits Kepler's meaning better, even though it does miss his verbal play on *Sanctum Officium* (see note 31, below).

30. In *De Revolutionibus,* in his dedicatory letter to Pope Paul III, Copernicus also mentions Lactantius as a revered theologian whose cosmological opinions are acknowledged to be false. See Lactantius, *Institut. Divin.,* III. 24, and Augustine, *The City of God,* XVI. 9.

31. Latin, *Officium,* the so-called "Holy Office," by which name the Inquisition was officially known. Kepler literally says, "the Office is Holy," referring to its name and not implying approval.

had taken two steps in my effort to reach this central goal: first, that I had discovered that the planetary eccentrics all intersect in the body of the sun, and second, that I had understood that in the theory of the earth there is an equant circle, and that its eccentricity is to be bisected.

The third step towards the physical explanation. The eccentricity of Mars's equant is to be precisely bisected.

Now we come to the third step, namely, that it has been demonstrated with certainty, by a comparison of the conclusions of Parts 2 and 4, that the eccentricity of Mars's equant is also to be precisely bisected, a fact long held in doubt by Brahe and Copernicus.[32]

Therefore, by induction extending to all the planets (carried out in Part 3 by way of anticipation), since there are (of course) no solid orbs, as Brahe demonstrated from the paths of comets, the body of the sun is the source of the power that drives all the planets around. Moreover, I have specified the manner [in which this occurs] as follows: that the sun, although it stays in one place, rotates as if on a lathe,[33] and out of itself sends into the space of the world an immaterial *species*[34] of its body, analogous to the immaterial *species* of its light. This *species* itself, as a consequence of the rotation of the solar body, also rotates like a very rapid whirlpool throughout the whole breadth of the world, and carries the bodies of the planets along with itself in a gyre, its grasp stronger or weaker according to the greater density or rarity it acquires through the law governing its diffusion.

Once this common power was proposed, by which all the planets, each in its own circle, are driven around the sun, the next step in my argument was to give each of the planets its own mover, seated in the planet's globe (you will recall that, following Brahe's opinion, I

32. For these terms, see the Glossary. Kepler means that the eccentricity of the equant point is twice as great as the eccentricity of the eccentric.

33. It should be pointed out that this remarkable conjecture was written several years before Galileo's observation of the sun's rotation.

34. This word can mean form, image, kind, emanation, spectacle, outward appearance, apparition, or idea (to name a few possibilities). Since there is no English word that can embrace so many meanings, I have chosen to leave it untranslated. See the Glossary for more about this term.

had already rejected solid orbs). And this, too, I have accomplished in Part 3.

By this train of argument, the existence of the movers was established. The amount of work they occasioned me in Part 4 is incredible, when, in producing the planet-sun distances and the eccentric equations[35] that were required, the results came out full of flaws and in disagreement with the observations. This is not because they should not have been introduced, but because I had bound them to the millstones (as it were) of circularity, under the spell of common opinion. Restrained by such fetters, the movers could not do their work.

Fourth step to the physical explanation. The planet describes an oval path.

But my exhausting task was not complete: I had a fourth step yet to make towards the physical hypotheses. By most laborious proofs and by computations on a very large number of observations, I discovered that the course of a planet in the heavens is not a circle, but an oval path, perfectly elliptical.

Geometry gave assent to this, and taught that such a path will result if we assign to the planet's own movers the task of making the planet's body reciprocate[36] along a straight line extended towards the sun. Not only this, but also the correct eccentric equations, agreeing with the observations, resulted from such a reciprocation.

Finally, the pediment was added to the structure, and proven geometrically: that it is in the order of things for such a reciprocation to be the result of a magnetic corporeal faculty. Consequently, these movers belonging to the planets individually are shown with great probability to be nothing but properties of the planetary bodies themselves, like the magnet's property of seeking the pole and catching up iron. As a result, every detail of the celestial motions is caused and regulated by faculties of a purely corporeal nature, that is, magnetic,

35. The mathematical expressions giving the planet's angular positions around the sun with respect to the apsides at specified times. The distances and the eccentric equations together are sufficient to specify the planet's position in space. Kepler's chief difficulty was that he could not find a single theory that could give both of these correctly. The solution, as he implies here, was to abandon the circular orbit and to introduce an ellipse.

36. This is explained in Chapter 57, below.

with the sole exception of the whirling of the solar body as it remains fixed in its space. For this, a vital faculty[37] seems required.

Next, in Part 5, it was demonstrated that the physical hypotheses we just introduced also give a satisfactory account of the latitudes.

There are some, however, who are put off by a few extraneous and seemingly valid objections and do not wish to put such great trust in the nature of bodies. Therefore, in Parts 3 and 4, some room was left for Mind, so that the planet's proper mover could attach the faculty of Reason to the animate faculty[38] of moving its globe. These people would have to allow the mind to make use of the sun's apparent diameter as a measure of the reciprocation, and to be able to sense the angles that astronomers require.

[The final paragraph of the Introduction, describing the Synoptic Table and the Summaries of the Chapters that follow, is omitted.]

---

37. Latin, *facultas vitalis.* In animal physiology, the faculty that facilitates bodily motions (other than motion of the body as a whole from place to place). See note 5, above.

38. Latin, *facultas animalis.* See footnote 5 above.

Chapter 1
On the distinction between the first motion and the second or proper motions; and in the proper motions, between the first and second inequality

Kepler begins the *Astronomia Nova* with a very general treatment of the geometrical hypotheses and models that astronomers customarily used. In doing this, he had two purposes.

First, he wanted to consider the equivalence or nonequivalence of different models. Although this had been done by both Ptolemy and Copernicus, Kepler goes into the subject much more deeply. In particular, he had found that slight changes in the models, though seeming innocuous at first, could produce errors in surprising ways. This general treatment prepares us for his demonstration (which we will only touch upon in this book) that all previous astronomers had gotten the earth's orbit (the sun's orbit, in Ptolemy) wrong. We will consider this when we come to Chapter 32.

Second, he wanted to introduce the question, central to the *Astronomia Nova* (and to the present selection), of what it is that makes the planets move. Most astronomers at that time had come to doubt that the planets are really transported by invisible spherical mechanisms. In the present chapter, Kepler points out the consequences of the removal of the spheres: the planets must somehow move through "the aethereal air," as he calls it. The second diagram in this chapter, a momentous step in planetary theory, shows the path that Mars would have to follow if Ptolemy and Aristotle were correct and the earth stood still. The result of a combination of regular circular motions turns out to be anything but perfect!

The following is Kepler's text.

The testimony of the ages confirms that the motions of the planets are orbicular. It is an immediate presumption of reason, reflected in experience, that their gyrations are perfect circles. For among figures it is circles, and among bodies the heavens, that are considered the most perfect. However, when experience is seen to teach something different to those who pay careful attention, namely, that the planets deviate from a simple circular path, it gives rise to a powerful sense of wonder, which at length drives people to look into causes.

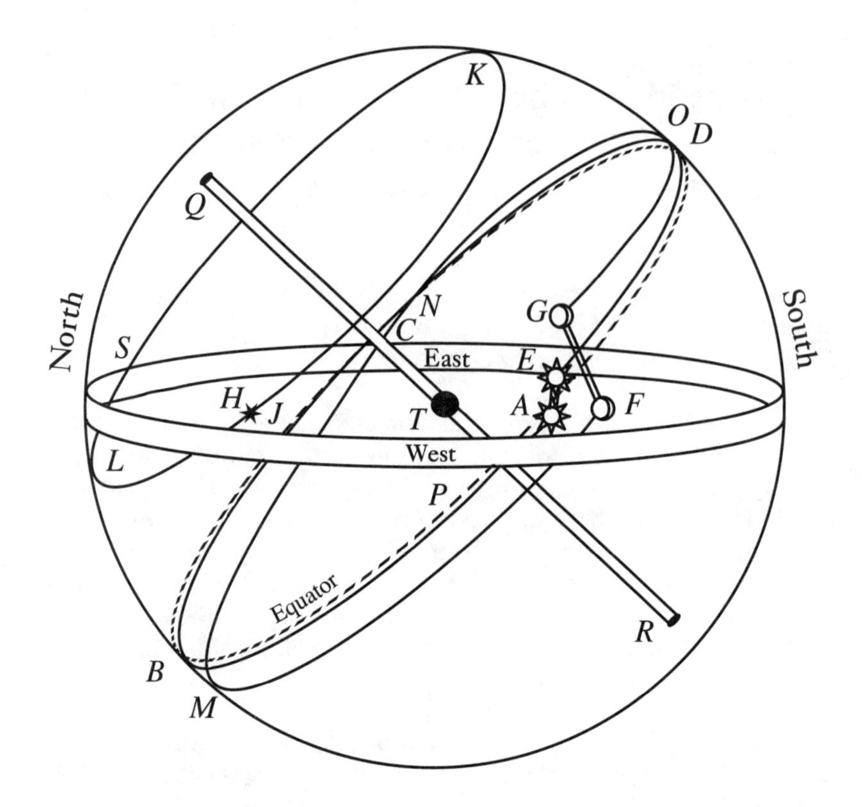

It is just this from which astronomy arose among humans. Astronomy's aim is considered to be to show why the stars' motions appear to be irregular on earth, despite their being exceedingly well ordered in heaven, and to investigate the circles wherein the stars may be moved, that their positions and appearances at any given time may thereby be predicted.

Here Kepler enters on a lengthy and rather abstruse account of the motions of the heavens, referring to the above diagram. His account may be summarized as follows.

Imagine the earth at *T*, surrounded by a sphere on whose surface the sun, moon, stars, and planets may be imagined to be located. Suppose we are viewing the sky at sunset, and we see a very bright star at *H* setting along with the sun (at *A*). This star will go below the horizon, move on to *L*, and then will rise again at *S* and proceed past *K* to *J*, which is the same point as *H*. This path is a circle on the imaginary celestial sphere.

During the same time in which the star traverses *HLSKJ*, the sun, setting at *A* (at this moment, just south of the equator), goes past *B*, *N*, and *D*, moving gradually northward as it circles around, and ends up at *E*, a little short of *A*.

The moon falls even farther behind. If it began at *F*, setting along with the sun, it would only get as far as *G* when the star arrives at *J*. It also can move farther northward or southward than the sun does in the same time; in this example, it is depicted as moving northward.

Neither of the motions of the sun and moon ever quite return to the same path. Instead, as Kepler puts it, they are "entwined one upon another like yarn on a ball." The motion, he says, "cannot be explained in figures or numbers, nor can it be extrapolated into the future, since it is always different from itself, to the extent that no spiral is equal to any other in elapsed time, and none carries over into the next with a curvature of the same quantity."

It was thus very helpful to astronomers to understand that two simple motions, the first one and the second ones, the common and the proper,* are intermingled, and that from this mingling that continuous sequence of conglomerated motions necessarily follows. And so, when that common and extrinsically derived diurnal revolving effect is removed, the fixed stars are suddenly no longer the swiftest and the moon slowest, but quite the opposite. ...

This turned out to be of great profit in astronomy for grasping the simplicity of the motions. Instead of unending spirals, an entirely new one always being added to the end of the earlier one at *E* or *G*, there remained little but the solitary circles *FG* and *AE*,[1] and a single

* Terms:

1.The *first motion* is that of the whole heaven and of all its stars from the east past the meridian to the west, and from the west through the lowest part of the heavens to the east, in the period of 24 hours; in the present diagram, *ABCD*.

2. The *second motions* are those of the individual planets from the west to the east, from *A* to *E*, from *F* to *G*, in longer periods. [Kepler's note]

---

1. *AE* and *FG* are representations of the motions of the sun and moon, respectively, against the background of the fixed stars. They are not complete circles, but are, at least approximately, segments of circles. The sun's path lies on a great circle called the ecliptic, which also rotates with the

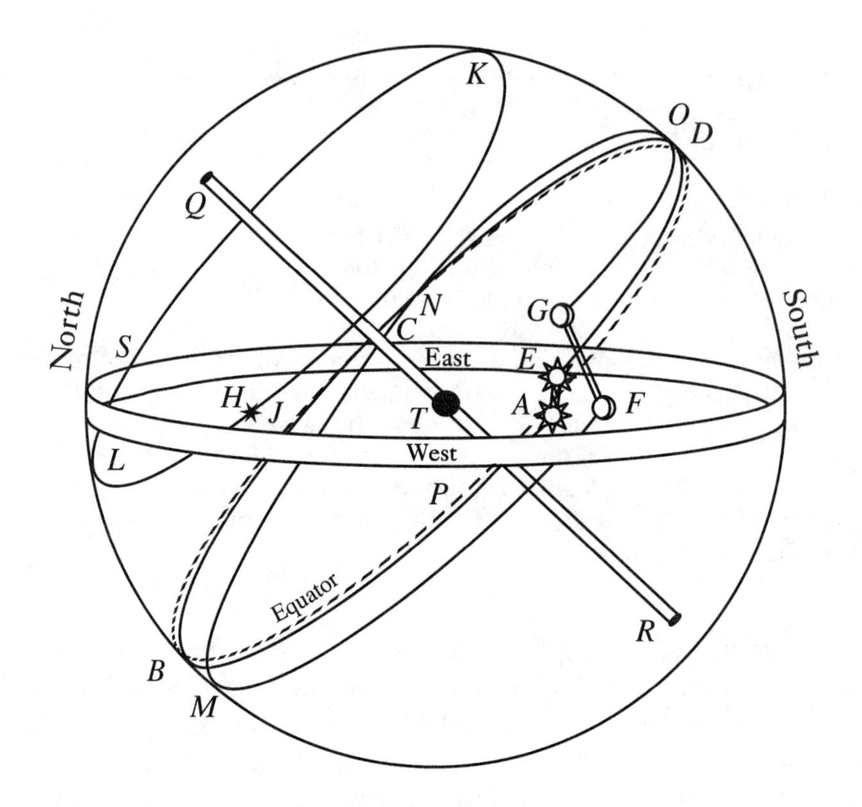

common [daily] motion, either of all the planets and the whole world as well in a direction opposite to the proper motions, or (following Aristarchus, making the world stand still) of the earth's globe *T*

(footnote continued)

common daily motion. The moon's path does not lie exactly on the ecliptic, but is within a few degrees of it.

One might suppose from Kepler's words that the circular nature of these paths is obvious. In fact, it isn't at all obvious. Because the sun can't be seen against the background of the stars, its path through the heavens must be deduced from other observations. And although the moon can be seen against the stars, its "circle" wobbles around so much as to raise the question of whether the circle found by the astronomers might not be merely an imaginary device.

However that may be, Kepler's point here is that it is much easier to find repeating patterns in the paths when the daily rotation is removed from consideration.

These motions are not true orbits; rather, they are projections of the actual orbits onto the surface of the imaginary celestial sphere.

around the axis *QR* in the same direction as the proper motions.

...

First, it was apparent that the three superior planets, Saturn, Jupiter, and Mars, attune their motions to their proximity to the sun. For when the sun approaches them they move forward and are swifter than usual, and when the sun comes to the sign opposite the planets they retrace with crablike steps the road they have just covered.[2] Between these two times they become stationary. These things always occur, no matter what the sign in which the planets are seen. At the same time, it was clear to the eye that the planets appear large when retrograde, and small when anticipating the coming of the sun with a swift and direct motion. From this, the conclusion was easily reached that when the sun approaches they are raised up and recede from the earth, and when the sun departs towards the opposite sign they descend again towards the earth. And finally, it was observed that this phenomenon of retrogression and increase of luminosity, just described, moves through the signs of the zodiac in order, tending from west through the meridian eastward, so that whatever has happened at one time in Pisces soon comes to pass similarly in Aries, then in Taurus, and so on in consequence.*

If one were to put all this together, and were at the same time to believe that the sun really moves through the zodiac in the space of a year, as Ptolemy and Tycho Brahe believed, he would then have to grant that the circuits of the three superior planets through the aethereal space, composed as they are of several motions, are real spirals, not (as before) in the manner of balled up yarn, with spirals set side by side, but more like the shape of pretzels,[3] as in the following diagram.

* Term.
In consequence means "according to the sequence of signs" (Aries, Taurus, and so on [see the diagram on the following page]), which series runs from the west through the meridian to eastern parts, then down towards the nadir and again to the west: from *F* to *G*, from *A* to *E*. [Kepler's note]

---

2. This "crablike"—i.e., backward—motion is called "retrogression" or "retrogradation."

3. The Latin is *panis quadragesimalis,* that is, "bread of the forty [days]," or lenten bread. Pretzels were invented by monks of southern Germany, who made it a practice to give them to children as treats during lent.

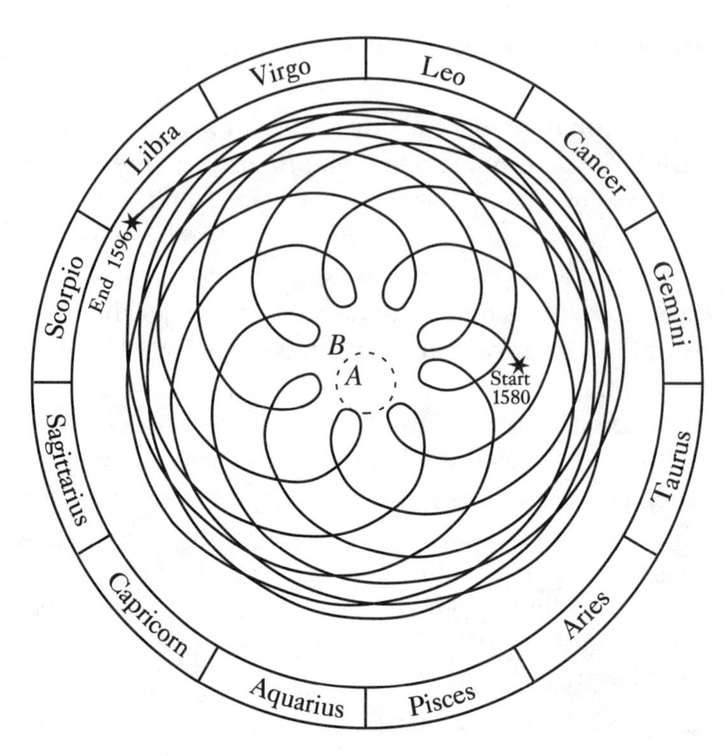

Kepler's description of the above diagram

This is the accurate depiction of the motions of the star Mars, which it traversed from the year 1580 until 1596, on the assumption that the earth stands still, as Ptolemy and Brahe would have it. These motions, continued farther, would become unintelligibly intricate, for the continuation is boundless, never returning to its previous path. Take note, too, that since the circle of Mars requires such a vast space, the spheres of the sun, Venus, Mercury, the moon, fire, air, water, and earth, have to be included in the tiny little circle around the earth *A*, and in its little area *B*. In addition, the greatest part even of this little space is given to Venus alone, much greater in proportion than is given to Mars here out of the whole area of the diagram. Moreover, we are forced to ascribe similar spirals to the remaining four planets if the earth stands still, and Venus's spiral would in fact be much more complicated.

Ptolemy and Brahe offer explanations of the causes, order, permanence, and regularity of these spirals, the former using individual epicycles carried around on the eccentrics of the individual planets, in imitation of the sun's motion, and the latter by having all the eccentrics carried around upon the single orb of the sun. Nevertheless, both leave the spirals themselves in the heavens. Copernicus, by attributing a single annual motion to the earth, entirely rids the planets of these extremely intricate spirals, leading the individual planets into their respective orbits, quite bare and very nearly circular. In the period of time shown in the diagram, Mars traverses one and the same orbit as many times as the "garlands" you see looped towards the center, with one extra, making nine times, while at the same time the earth repeats its circle sixteen times.

The figure on the facing page is a momentous diagram. Nothing like it had ever before been published. Astronomers had become so accustomed to thinking of celestial motions as compounded circular motions that it had apparently not occurred to anyone to consider the actual path traversed by a planet. Once the reality of the celestial spheres came into question, however, the actual path traversed came to be a matter of great interest.

Throughout the *Astronomia Nova*, we will see Kepler gaining new insight into the celestial motions by conceiving of them as occurring in three-dimensional space rather than on the surface of the imaginary celestial sphere.

Again, however, it was noticed that these loops in each planet's spirals are unequal in different signs of the zodiac, so that in some places the planet would retrogress through a longer arc of the zodiac, at others through a shorter, and now for a longer, now for a shorter time. Nor is the increment of brightness of a retrograde planet always the same. Also, if one were to compute the times and distances between the midpoints of the retrogressions, neither times nor arcs would be equal, nor would any of the times answer to its arc in the same proportion. Nevertheless, for each planet there was a certain sign of the zodiac from which, through the semicircle to the opposite sign in either direction, all those things successively increased.

From these observations it came to be understood that for any planet there are two inequalities[4] mixed together into one, the first of which completes its cycle with the planet's return to the same sign of the zodiac, the other with the sun's return to the planet.

Now the causes and measures of these inequalities could not be investigated without separating the mixed inequalities and looking into each one by itself. They therefore thought they should begin with the first inequality, it being more nearly constant and simple, since they saw an example of it in the sun's motion, without the interference of the other inequality.* But in order to separate the second

* The sun has only a single inequality, with respect to the time within which it is completed. But as for the causes of this inequality, the same two factors combine as much for the sun as for the other planets, as will be explained below. [Kepler's note]

4. For definitions of the two inequalities, see the Glossary.

inequality from this first one, they could proceed no otherwise than by considering the planets on those nights at whose beginning they rise while the sun is setting, which thence were called *akronychioi*, or night rising.[5] For since the presence and conjunction of the sun makes them go faster than usual, and the opposition of the sun has the opposite effect, before and after these points they are surely much removed from the positions they were going to occupy through the action of the first inequality. Therefore, at the very moments of conjunction with and opposition to the sun they are traversing their own true and proper positons. But since they cannot be seen when in conjunction with the sun, only the opposition to the sun remains as suitable for this purpose.

But since the sun's mean and apparent motions* are two different things, for the sun, too, is subject to the first inequality, the question is raised which of these removes the second inequality from the planet, and whether the planets should be considered when at opposition to the sun's apparent position or its mean position. Ptolemy chose the mean motion, thinking that the difference (if any) between taking the mean sun and the apparent sun could not be perceived in the observations, but that the form of computation and of the proofs would be easier if the sun's mean motion were taken. Copernicus and Tycho followed Ptolemy, carrying over his assumptions. I, as you see in Chapter 15 of my *Mysterium cosmographicum,* take instead the apparent position, the true body of the sun, as my reference point, and will vindicate that position with proofs in Parts 4 and 5 of this work.

But before that, I shall prove in this first part that one who substitutes the sun's apparent for its mean motion sets up a completely different orbit for the planet in the aether, whichever of the more celebrated opinions of the world he follows.[6] Since this proof depends upon the equivalence of hypotheses, we shall begin with this equivalence.

* Terms:

The sun's *apparent position* is that which it is perceived to occupy through its inequality.

The *mean position* is that which it would have occupied if it had not had its inequality. [Kepler's note]

---

5. This applies only to the superior planets (Mars, Jupiter, and Saturn). Other methods had to be applied to Venus and Mercury.

6. The demonstration of this is carried out in Chapter 6 (not part of the present selection).

Chapter 2
On the first and simple equivalence, that of the eccentric and the concentric with an epicycle, and their physical causes

In this chapter, Kepler continues the themes of Chapter 1. We will skip the proof of equivalence (since, after presenting the diagrams, Kepler simply refers the reader to Ptolemy), and take up the perplexing question of how a planet could be made to move on an eccentric circle either by physical forces or by an animate or intelligent moving power.

What Kepler finds is that it is remarkably difficult to generate the supposedly simple and perfect regular circular motion. The form of the circle must in some way be there in space already for the planet to follow; otherwise, even if it is moved by an intelligence, it doesn't have enough information to generate the circle.

Newton revisits this question in *Principia* Book I Proposition 7, and finds essentially the same thing: the planet's velocity at any point on the circle depends on where it will be when it gets around to the other side of the center of forces. So it seems that there is something awkward and *unnatural* about "simple" circular motion!

Kepler is, of course, preparing us to accept the idea of an oval or elliptical orbit, which he first introduces in Chapter 44.

And now, to begin, I take up the equivalence of hypotheses adopted to save the appearances of the first inequality, which were demonstrated by Ptolemy in Book III and by Copernicus in Book III Chapter 15.

We will omit Kepler's discussion of the geometrical and physical comparison of different arrangements of spheres. Kepler did not believe they existed, since careful observations had shown that comets are celestial objects, and would therefore have to pass through the supposed spheres. This raises the question, "how do the planets know how to move?"

But, with arguments of the greatest certainty, Tycho Brahe has demolished the solidity of the orbs, which hitherto was able to serve these moving souls, blind as they were, as walking sticks for finding their appointed road. This entails that the planets complete their courses in the pure aether, just like birds in the air. Therefore, we shall have to philosophize differently about these models.

Let it then be taken as a first principle, that each force by which motions of this sort are administered dwells in the body of the planet itself, and is not to be sought outside it.

Now the planet must execute a perfectly circular path in the pure aether by its inherent force.... It is therefore clear that the mover is going to have two jobs: first, it must have a faculty strong enough to move its body about, and second, it must have sufficient knowledge to find a circular boundary in the pure aether, which in itself is not divided into such regions. This is the function of mind. Please don't tell me that the motive faculty itself, as a member of the family of simple and brute souls, has a native aptitude for circular motion, exactly like a stone's nature to descend in a straight line. For I deny that God has created any perpetual non-rectilinear motion that is not ruled by a mind. Even in the human body, all the muscles move according to the principles of rectilinear motions. They either swell by contracting into themselves, so that the member approaches the muscle, or stretch out, the ends moving apart, so that the member recedes. This takes place similarly, though in its own way, in the circular muscles which are set up in passages as guards. When they are extended by the circular filaments, they relax and open the passage, and constrict it when the filaments contract into the form of a smaller circle. There is no member whatever that rotates uniformly and without impediment. The bending of the head, feet, arms, and tongue is expressed in certain mechanical devices by the contraction of many straight muscles carried across from one place to another. In this way it is brought about that the motive faculty, which by its own nature tends in a straight line, swings its member in a gyre. Likewise, certain machines raise water to great heights, not because the nature of the body, which produces its motion, tends to an exalted position, but because, by an arrangement of channels, it is brought about that the water necessarily gives way upwards when a greater weight tends downwards. And even if the motion of certain members were perfectly circular, it nonetheless could not be perpetual. There should be no great wonder at this, since in the human body mind presides over the animate faculty. Surely, then, if there had been any way of so constructing some moving faculty that some body might be able to rotate, it would not have been neglected in the human body.

Besides, it is quite impossible for any mind to manifest a circular path without the guidepost either of a center or of some body which might appear under a greater or smaller angle according to its approach or recession. For a circle is both defined and brought to perfection by the same criterion, namely, equality of distance from the middle. No matter how many of these motive faculties you set up, a circle, even for God, is nothing but what was just said. Geometers do, of course, show how, given three points on a circumference, to form a continuous circle, but this itself presupposes that some portion of the circumference (that which passes through the three points) is already constructed. Who, then, will show the planet this starting place, in conformity with which it will make the rest of its path? This is impossible unless the planet's mover (as in Avicenna's opinion) imagines for itself the center of its orb and its distance from it, or is assisted by some other distinguishing property of a circle in order to lay out its own circle.

Kepler then shows that numerous absurdities result from the supposition that the moving power somehow emulates one of the two basic circular forms of hypotheses. He concludes:

I have presented these models hypothetically, the hypothesis being astronomy's testimony that the planet's path is a perfect eccentric circle such as was described. If astronomy should discover something different, the physical theories will also change.

The equivalence of these hypotheses lies not so much in the equality of the apparent angles [around the center] as in the identity of the actual paths of the planets through the surrounding aether. ...

Kepler concludes the chapter here with a sentence referring to the specific models considered earlier. Again, as in Chapter 1, he emphasizes the importance of the planet's actual path, as distinct from the mechanisms that are supposed to generate it.

Chapter 7
The circumstances under which I happened upon the theory of Mars

It is true that a divine voice, which enjoins humans to study astronomy, is expressed in the world itself, not in words or syllables, but in things themselves and in the conformity of the human intellect and senses with the sequence of celestial bodies and of their dispositions. Nevertheless, there is also a kind of fate, by whose invisible agency various individuals are driven to take up various arts, which makes them certain that, just as they are a part of the work of creation, they likewise also partake to some extent in divine providence.

When, in my early years, I was able to taste the sweetness of philosophy, I embraced the whole of it with an overwhelming desire, and with no special interest whatever in astronomy. I certainly had enough intelligence, nor did I have any difficulty understanding the geometrical and astronomical topics included in the normal curriculum, aided as I was by figures, numbers, and proportions. These were, however, required courses, and did not suggest a particular inclination for astronomy. And since I was supported at the expense of the Duke of Württemberg, and saw my comrades, whom the Prince, upon request, was sending to foreign countries, stalling in various ways out of love for their country, I, being hardier, quite maturely agreed with myself that whithersoever I was destined I would promptly go.

The first to offer itself was an astronomical position; however, to tell the truth, I was driven to take on this task by the authority of my teachers. I was not frightened by the distance of the place, for (as I have just said) I had condemned this fear in others, but by the low opinion and contempt in which this kind of function is held, and the sparsity of erudition in this part of philosophy. I therefore entered upon this better furnished with wits than with knowledge, and protesting loudly that I would never willingly concede my intention to follow another kind of life which seemed more splendid.[1] What came of those first two years of study may be seen in my *Mysterium cosmographicum*. The additional goads which my teacher Maestlin gave me towards embracing the rest of astronomy, you will read of in the same book, and in that man's prefatory letter to Rheticus's

1. Kepler had hoped eventually to obtain an ecclesiastical post.

Narratio.[2] I had the very highest opinion of the discovery,[3] and all the more so when I saw that Maestlin, too, held it in similar esteem. It was not so much his untimely promise to the readers of what he called a "uranic opus" of my universe that spurred me on, as it was my own ardor to seek, through a reworking of astronomy, whether my discovery would stand comparison with observations made with perfect accuracy. For it had then been demonstrated in that book that the discovery was consistent with the observations within the limits of accuracy of ordinary astronomy.

So from that time I began to think seriously of comparing observations. In 1597, I wrote Tycho Brahe asking his opinion of my little book,[4] and when he, in reply, mentioned among other things his own observations, he ignited in me an overwhelming desire to see them. Moreover, Tycho, who was indeed himself a large part of my destiny, did not cease from then on to invite me to come to him. And since I was frightened off by the distance of the place, I again ascribe it to divine arrangement that he came to Bohemia. I thereupon came to visit him at the beginning of 1600 in hopes of learning the correct eccentricities of the planets. But when I found out during the first week that, like Ptolemy and Copernicus, he made use of the sun's mean motion, while the apparent motion would be more in accord with my little book (as is clear from the book itself), I begged the master to allow me to make use of the observations in my own manner. At that time, the work which his aide Christian Severinus[5] had in hand

---

2. Kepler's mathematics professor at Tübingen, Michael Maestlin, who saw to the printing of the *Mysterium cosmographicum*, took the liberty of appending Georg Joachim Rheticus's *Narratio prima* (1540) to aid the reader in understanding the Copernican hypothesis.

3. Kepler's *Mysterium Cosmographicum* (1596) showed that the distances of the six Copernican planets (Mercury, Venus, Earth, Mars, Jupiter, and Saturn) from the sun could with fairly good accuracy be determined by a concentric arrangement of the five regular geometric solids (octahedron, icosahedron, dodecahedron, tetrahedron, and cube, in that order) inscribed in, and circumscribed about, spherical surfaces. He believed that he had discovered the reason why God had created exactly six planets, and why the Creator had arranged them as He did.

4. Letter to Tycho Brahe, 13 December 1597, letter number 82 in *KGW* 13 p. 154.

5. Better known under the name "Longomontanus," the Latinization of his birthplace, Longberg.

was the theory of Mars. The occasion had placed this in his hands, in that they were busy with the observation of the acronychal position or opposition of Mars to the sun in 9° Leo. Had Christian been treating a different planet, I would have started on it as well.

I therefore once again think it to have happened by divine arrangement, that I arrived at the same time in which he was intent upon Mars, whose motions provide the only possible access to the hidden secrets of astronomy, without which we would remain forever ignorant of those secrets.[6]

A table of mean oppositions was worked out, starting with the year 1580. A hypothesis was invented which, it was proclaimed, represented all these oppositions within a distance of two minutes in longitude. ... It was only in the latitude at acronychal positions and also the parallax of the annual orb[7] that Christian got stuck. There was, actually, a hypothesis and table for the latitudes, but they failed to elicit the observed latitude. This same result was destined to be a problem for the lunar theory as well.

Now since I suspected what proved to be true, that the hypothesis was inadequate, I entered upon the work girded with the preconceived opinions expressed in my *Mysterium cosmographicum*. At the beginning there was great controversy between us as to whether it were possible to set up another sort of hypothesis which would express to a hair's breadth so many positions of the planet, and whether it were possible for the former hypothesis to be false despite its having accomplished this so far over the entire circuit of the zodiac.

I consequently showed, using the arguments presented already in Part I, that an eccentric can be false, yet answer for the appearances within five minutes or better, provided that the center of the equant be correctly known. As for the parallax of the annual orb, and the latitudes, that prize is not yet won, and besides, was not attained by their hypothesis. What remains, then, is to find out whether they, with their means of computation, might not somewhere differ from the observations by five minutes.

---

6. Mars and Mercury are the only planets whose orbits differ enough from circles for that difference to have an effect observable by Tycho's instruments. Mercury is too near the sun to afford reliable observations of its entire orbit. Therefore, only the observations of Mars could have led Kepler to his "new astronomy."

7. That is, the second inequality, resulting from the earth's motion.

I therefore began to investigate the certitude of their operation. What success came of that labor, it would be boring and pointless to recount. I shall describe only so much of that labor of four years as will pertain to our methodical enquiry.

Chapter 24
A more evident proof that the epicycle or annual orb is eccentric with respect to the point of uniformity

To prepare the reader for accepting his "new astronomy founded on physics" (as he put it on the title page), Kepler needed to prove two main points. The first of these was that a standard, Ptolemaic-style theory of the first inequality, consisting of an eccentric circular deferent with a fixed equant point, couldn't be made to work. No matter how the eccentricity of the deferent and the equant were adjusted, there was always some discrepancy between the theory and the observations it was meant to account for. This demonstration is set forth in Chapters 8–21. In Chapter 19, after noting the discrepancy, he remarks,

> Something among those things we have assumed must be false. But what was assumed was: that the orbit upon which the planet moves is a perfect circle, and that there exists some unique point on the line of apsides at a fixed and constant distance from the center of the eccentric about which point Mars describes equal angles in equal times. Therefore, of these, one or the other, or perhaps both, are false, for the observations used are not false. (*Astronomia Nova,* 1609, p. 112.)

The second thing Kepler needed to establish was that the theory of the earth (or of the sun, for Brahe) requires an equant; that is, that a simple eccentric circle with uniform motion around its center cannot adequately account for the phenomena. This is in a way surprising, because ever since the time of Hipparchus (second century BCE) such a theory had been used with no problems, and even Tycho Brahe had approved it on the basis of his very accurate observations.

While not disputing Brahe's reasoning, Kepler had for a long time been puzzled that the earth was the only planet in the Copernican system that moved uniformly on its circle. He believed on physical grounds that this could not be right, and sought a way of investigating earth's orbit more accurately.

Kepler noted that all theorists from Hipparchus through Brahe had constructed their theories solely on the basis of observations of the sun's apparent angular motion in the sky. Therefore, although such theories might be accurate in representing *angular* positions, they might be wrong about the earth's *distances* from the sun. But how could those distances be measured?

Kepler's ingenious solution to the problem involved, in effect, observing the earth's motion from Mars. If Mars stood still in the heavens, it could be used as a sort of survey marker by which people on earth could find their position, triangulating from the sun and Mars. Now of course Mars doesn't stand still, but it does return to the same place in slightly less than 687 days. This is the key to Kepler's examination of the earth's orbit. After examination of some special cases (in Chapters 22 and 23), Kepler takes up the general application of his triangulation method here in Chapter 24.

Although there is some mathematics in this chapter, those not interested in the details of the triangulation can just skip to the end, where the lengths of the lines *af, ah, ae,* and *ac* (distances of the earth from the center of regular motion *a*) are given. Here he concludes that *a* is not the center of the orbit, which is the main point he is trying to make in this chapter.

Such, then, was the beginning of this enquiry,[1] timid and encumbered with such concern that [the stated conditions be satisfied].

Now that we have once hazarded this, we are buoyed by audacity to sally forth again more freely onto the battlefield. For I shall seek out three or more observed positions of Mars with the planet always at the same eccentric position, and from these find by trigonometry the distances of that number of points on the epicycle or annual orb from the point of uniform motion. ...

The first time shall be 1590 March 5 at 7^h 10^m in the evening, since then Mars had hardly any latitude, and thus no one looking at the demonstration could be troubled by any irrelevant suspicions about the intermingling of latitude.[2] To this there correspond these moments, in which Mars returns to the same sidereal position: 1592 Jan. 21 at 6^h 41^m; 1593 Dec. 8 at 6^h 12^m; 1595 Oct. 26 at 5^h 44^m.[3] ...The true longitude in 1590 will be 15° 53' 45" Taurus, and for subsequent times 1' 36" greater for each. For this is the motion of precession

1. This is a reference to the somewhat limited investigation presented in the two preceding chapters.

2. When Mars has no latitude (see the Glossary for latitude), it lies in the plane of the earth's orbit. In the diagram below, the point *k,* representing Mars, is in the plane of the paper, not above or below it, and so no error can arise from points in the diagram not being in the same plane.

3. Kepler uses Mars's sidereal period of 686 days 23 hours and 31 minutes to determine the subsequent times.

corresponding to the periodic time of one return of Mars.[4]

. . .

We shall present this first in the Copernican form, since it is simpler to perceive.

Here Kepler begins the geometrical demonstration, which he distinguishes from the more discursive parts of the chapter by the use of italic type. For the sake of clarity, we will preserve Kepler's distinction by indenting the mathematics while keeping the text in the usual typeface.

Let a be the point of uniform motion of the earth's circuit, which shall be considered to be the circle dg described about a,[5] and let the sun be on the side b, such that the line of the sun's apogee ab lies in the direction of 5½° Cancer,[6] despite our being about to investigate this freely, as if unknown, in Chapter 25.

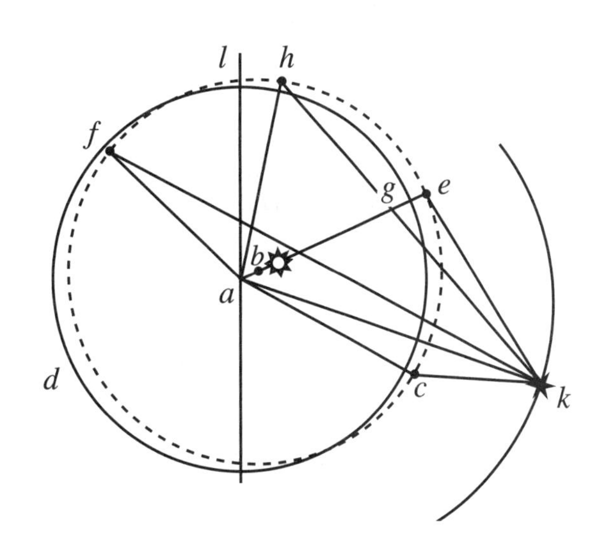

4. The line *ak* is motionless with respect to the fixed stars. However, its position is measured with respect to the spring equinox point, which slowly drifts backward against the background of the stars. Therefore, with respect to the spring equinox point, it is 1' 36" farther forward for each revolution of Mars. The appropriate multiple of this amount must be added to the position of *ak* in 1590 (15° 53' 45" Taurus), which is obtained from Brahe's Mars theory.

5. The solid-line circle *dg* is the earth's path as it would have appeared in a standard Copernican theory (updated using Brahe's parameters). It is to be "considered" the correct path until proved otherwise below.

6. This is taken from Brahe's solar theory. In the diagram, the earth's line of apsides coincides with the position of line *ae*. In fact (as may be deduced from the numbers below) *ae* is at about 27° Gemini, about eight degrees before the perihelion. However, since the position of the earth's perihelion is never used in this demonstration, this difference is not important, and Kepler has simplified the diagram by making the sun fall on *ae*.

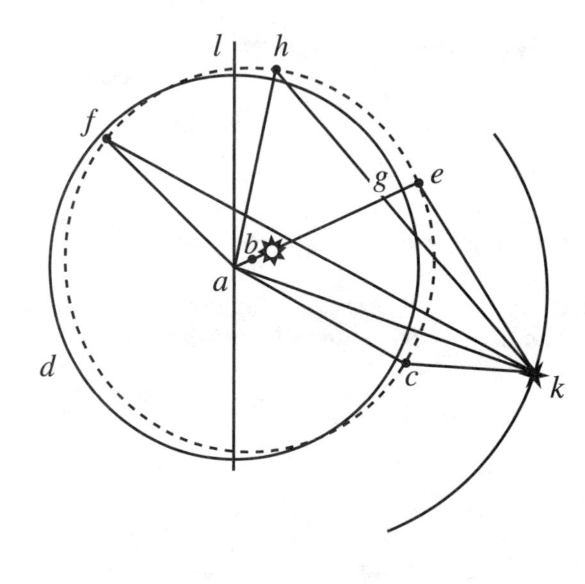

And let the earth be on the line *af* in 1590, *ah* in 1592, *ae* in 1593, and *ac* in 1595.

And the angles *fah, hae, eac* are equal,
since *a* is the point of uniform motion
and the periodic times of Mars are presupposed equal.

And let the planet at these four times be at *k*, and its line of apsides be *al* [at 23½° Leo].

The angle *fak*…is 127° 5' 1".[7]

As for the observed position of Mars, at the same time on the preceding day, the fourth, it was at 24° 22' Aries.

Its diurnal motion for the day would be 44'.

Therefore, at our time it was seen at 25° 6' Aries,
which is the position of the line *fk*.

But *ak* is directed towards 15° 53' 45" Taurus [from p. 46].

Therefore, *fka* is 20° 47' 45". So the remainder *afk* to make up two right angles is 32° 7' 14".[8]

Now as the sine of *afk* is to *ak* , which we shall say is 100,000 units, so is [the sine of] *fka* to *fa,* which is what is sought. Therefore, *fa* is 66,774.[9]

---

7. This is the angle between the positions of earth and Mars as measured around the center of the earth's orbit. It is derived from Brahe's planetary models. In this chapter, Kepler is showing that even Brahe's own theories convict his solar theory of error. In subsequent chapters, Kepler carries out similar computations using only observed positions, in an elaborate procedure that is much more careful and much more accurate.

8. The angle *fka* is found by subtracting 25° 6' Aries from 15° 53' 45" Taurus. Since the sum of the angles of any triangle is equal to two right angles (by Euclid I.32), angle *afk* = 180° – (angle *fak* + angle *fka*).

9. This is by the law of sines, which states that in any triangle the sine of each

Now if the remaining lines *ha, ea, ca,* turn out to be of the same length, my suspicion will be false, but if they are different, my triumph will be complete.

Second, then, in 1592 at our moment...
the angle *hak* is 84° 10' 34".[10]

It was observed on January 23 at 7^h 15^m at 11° 34½' Aries,...

And its motion over two days was 1° 25'.

Therefore on the 21st at 7^h 15^m it was seen at 10° 9½' Aries [line *hk*].

The remaining parts of an hour would deduct the half minute.

[In 1592, *ak* is directed towards 15° 55' 21" Taurus.]

Therefore, the angle *hka* is 35° 46' 23",
and *ahk* is 60° 3' 3",
and *ah* is 67,467, now longer than *af*.

This is doubtless because the sun has descended towards perigee, and the earth has been moved from *f* to *h*; thus, in this region the sun is found beyond *b* at a nearer point.[11]

Third, in 1593 at our moment...*eak* was 41° 16' 16".

It was observed December 10 at 7^h 20^m at 4° 45' Aries....

Its motion over two days was 1° 8'.

Therefore, on December 8 at 7^h 20^m it was seen at 3° 37' Aries,
while at our time of 6^h 12^m it was at 3° 35½' Aries [line *ek*].

[In 1593, *ak* is directed towards 15° 56' 57" Taurus.]

Hence, *eka* is 42° 21' 30"
and *kea* is 96° 22' 14",
and *ae* is 67,794,
again longer, for it is yet closer to the sun's perigee.

(footnote continued)

angle divided by the opposite side is the same for all three angles.

10. Again derived from Brahe's models.

11. This remark is rather obscure. What Kepler may mean is that if the earth is not moving on the circle *dg* around *a*, but on another circle (dashed line) whose center is closer to the sun, then as the earth gets closer to perihelion (or the sun closer to perigee) the distance *ah* is going to be greater.

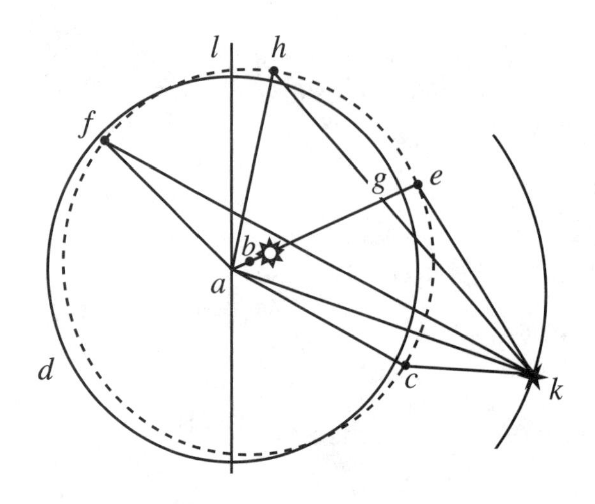

Fourth, in 1595 at our moment...the angle *kac* was 1° 38' 5".

It was observed on October 27 at 12^h 20^m at 18° 52' 15" Taurus, retrograde.[12]

Its diurnal motion was 23'.

And so on the 26th at 12^h 20^m it was at 19° 15' 15" Taurus,

while at our time it was at 19° 21' 35" Taurus [line *ck*].

[In 1595, *ak* is directed towards 15° 58' 33" Taurus.]

Therefore, *akc* is 3° 23' 5",

and the supplement of *ack* is 5° 1' 10",

and *ac* is 67,478.

End of the geometrical investigation.

...

I shall now present all four lines for inspection.

Sun's mean position	22° 59' Pisces	*af*	66,774
	10 6 Aquarius	*ah*	67,467
	27 13 Sagittarius	*ae*	67,794
	14 20 Scorpio	*ac*	67,478

So the longest is *ae*, which is also the closest to the sun's perigee; the shortest is *af*, which is also the farthest from the sun's perigee; and *ac* and *ah* are about equal, because they are also nearly equally removed from perigee.

Moreover, even if *ac* is a little longer than *ah* which is nearer the perigee, this should be attributed to the smallness of the angles at *c*, through which such a small error as this could easily be introduced. Therefore, the circle *dg*, which Copernicus described about the point *a* of uniformity of the earth's motion, is not the earth's path. There is instead some other circle *fhec* on which the earth is found, whose center lies in the same direction as the sun—that is, at *b*.

12. Because Mars was going backward, the diurnal motion must be subtracted.

Chapter 32
The power that moves the planet in a circle diminishes with removal from its source

We have just seen (in Chapter 24) one of the ways in which Kepler showed that the earth's orbit must have an eccentricity half that of the classical earth/sun models, with a separate equant point, exactly like the models for all the other planets in the sun-centered Copernican universe. Because this point was crucial to his introduction of physical principles to astronomy, Kepler proved it in many different ways, in Chapters 22–28. Chapters 29 and 30 present the computation of new earth-sun distances.

The question then naturally arose whether the new earth/sun model would give correct angular positions for the sun. What he found was that the maximum difference between Brahe's solar theory and the new one is only seven arc seconds—a completely negligible amount. This is shown in Chapter 31.

At this point, Kepler has proved two important things: first, that the old theories will not work, and second, that in the Copernican system all planets move according to the same principles. With this preparation, Kepler hoped that the reader may now be ready to consider his new physical account, which begins with the present chapter and concludes with the demonstration of the elliptical form of the orbit of Mars in Chapter 59.

It is worth noting that Kepler's new earth/sun theory did more to improve planetary tables than any other discovery he made, its effect being about an order of magnitude greater than that of his new speed law for the planets (the area law, or "Second Law") and of the elliptical form of the orbits.

I have said above that Ptolemy, well informed by the observations, bisected the eccentricities[1] of the three superior planets, that Copernicus imitated this, and that Tycho's observations of Mars urge the same conclusion, as has been seen in Chapters 19 and 20, and will appear with much greater certainty below in Chapter 42. In addition, Tycho closely

1. That is, the celestial body (or center of the epicycle) does not move uniformly around the center of its deferent circle, but instead moves uniformly around a point whose eccentricity is twice that of the deferent circle. For a full explanation of these terms, see Appendix A.

imitated this in his lunar theory. And now the same thing has been demonstrated in the theory of the sun (for Tycho) or of the earth (for Copernicus). Further, there is nothing to prevent our believing the same of Venus and Mercury. Indeed, I now have a proof that this is the origin of the belief that the centers of these planets' eccentrics move around on a small annual circle. Therefore all planets have this [eccentric circle with an equant]. Now in my *Mysterium cosmographicum*, published eight years ago,[2] I postponed arguing this case of the cause of the Ptolemaic equant for the sole reason that it could not be said on the basis of ordinary astronomy whether the sun or earth uses an equalizing point and has its eccentricity bisected. However, now that we have the confirmation of a sounder astronomy, it should be transparently clear that there is indeed an equant in the theory of the sun or earth. And, I say, now that this is demonstrated, it is proper to accept as true and legitimate the cause to which I assigned the Ptolemaic equant in the *Mysterium cosmographicum*, since it is universal and common to all the planets.[3] So in this part of the work I shall make a further declaration of that cause.

And since the declaration will be general, I shall use the word, "planet." However, in this and the next few chapters, the reader may always understand this as denoting in particular the earth for Copernicus or the sun for Tycho.

First, the reader should know that in all hypotheses constructed according to this Ptolemaic form, however great the eccentricity, the speed at perihelion and slowness at aphelion are very closely proportional to the lines drawn from the center of the world to the planet.

Here Kepler presents an elaborate proof demonstrating that Ptolemy's equant with bisected eccentricity makes a planet's speed increase in

2. In the margin, Kepler inserted the comment "Now more than that." Since the *Mysterium* had been published in the winter of 1596–7, this chapter must have been written in 1604 or 1605, possibly before the discovery of the elliptical form of the orbit.

3. In Chapter 22 of the *Mysterium*, Kepler wrote, "You now have the cause of the equant's being at a distance...from the center of the eccentric: namely, that the whole universe shall be filled with a soul that shall catch up any of the stars or comets it reaches, with such a swiftness as the distance of the place from the sun, and the strength of the power there, requires. Next, that each of the planets shall have its own soul, by whose rowing action the star might ascend to its [proper] distance."

direct proportion to its closeness to the sun, very nearly. This can be shown more simply as follows.

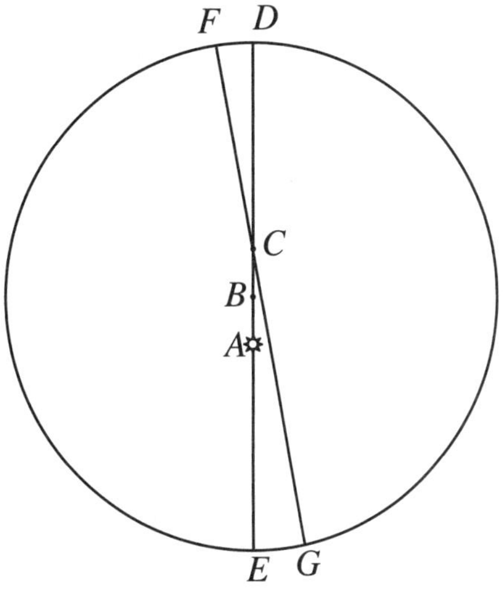

Let the eccentric be *DFEG*,
with center *B*,
aphelion *D*,
perihelion *E*,
the sun
(or earth,
for Ptolemy) *A*,
and the equant *C*.

And let $AB = BC$,
so as to give the hypothesis
the "Ptolemaic form."[4]

Through *C*
draw the straight line
FCG,
intersecting the circle
at *F* very near aphelion,
and at *G* very near perihelion.

Since *FCD* and *GCE* are equal,
and are angles about the equant point *C*,
they are traversed in equal times.

Because these angles are very small,
we may take it that side CD = side CF, approximately,
and that side CE = side CG, approximately,
and that the triangles *CDF* and *CEG* are very nearly similar.

Therefore,

$DF : GE :: CD : CE$.

But $AE = CD$, and $AD = CE$.

So $DF : GE :: AE : AD$

—that is, arcs traversed in equal times are inversely proportional to their distances from the sun.

4. Note that the demonstration depends on this "bisection of the eccentricity."

Chapter 33
The power that moves the planets resides in the body of the sun

It was demonstrated in the previous chapter that the elapsed times of a planet on equal parts of the eccentric circle (or on equal distances in the aethereal air) are in the same ratio as the distances of those spaces from the point whence the eccentricity is reckoned; or, more simply, to the extent that a planet is farther from the point which is taken as the center of the world, it is less strongly urged to move about that point. It is therefore necessary that the cause of this weakening is either in the very body of the planet, in a motive force placed therein, or right at the supposed center of the world.

Now it is an axiom in natural philosophy of the most common and general application that of those things which can occur at the same time and in the same manner, and which are always subject to like measurements, either one is the cause of the other or both are effects of the same cause. Just so, in this instance, the intensification and remission of motion is always in the same ratio as the approach and recession from the center of the world. Thus, either that weakening will be the cause of the star's motion away from the center of the world, or the motion away will be the cause of the weakening, or both will have some cause in common. But it would be impossible for anyone to think up some third concurrent thing which would be the cause of these two, and in the following chapters it will become clear that we have no need of feigning any such cause, since the two are sufficient in themselves.

Further, it is not in accord with nature that strength or weakness in longitudinal motion should be the cause of distance from the center. For distance from the center is prior both in thought and in nature to motion over an interval. Indeed, motion over an interval is never independent of distance from the center, since it requires a space in which to be performed, while distance from the center can be conceived without motion. Therefore, distance will be the cause of intensity of motion, and a greater or lesser distance will result in a greater or lesser amount of time.

And since distance is a kind of relation whose essence resides in end points, while of relation itself, without respect to end points, there can be no efficient cause, it therefore follows (as has been said) that the cause of the variation of intensity of motion inheres in one or the other of the end points.

Now the body of a planet is never by itself made heavier in receding, nor lighter in approaching.

Moreover, that an animate force, which the motion of the heavens suggests is seated in the mobile body of the planet, undergoes intensification and remission so many times without ever becoming tired or growing old—this will surely be absurd to say. Also, it is impossible to understand how this animate force could carry its body through the spaces of the world, since there are no solid orbs, as Tycho Brahe has proved. And on the other hand, a round body lacks such aids as wings or feet, by the moving of which the soul might carry its body through the aethereal air as birds do in the atmosphere, by some kind of pressure upon, and counter-pressure from, that air.

Therefore, the only remaining possibility is that the cause of this intensification and weakening resides in the other endpoint, namely, in that point which is taken to be the center of the world, from which the distances are measured.

The motive power is in the center of the system.

So now, if the distance of the center of the world from the body of a planet governs its slowness, and approach governs its speeding up, it is a necessary consequence that the source of motive power is at that supposed center of the world. And with this laid down, the manner in which the cause operates is also clear. For it gives us to understand that the planets are moved rather in the manner of the steelyard or lever. For if the planet is moved with greater difficulty (and hence more slowly) by the power at the center when it is farther from the center, it is just as if I had said that where the weight is farther from the fulcrum, it is thereby rendered heavier, not of itself, but by the power of the arm supporting it at that distance. And this is true, both of the steelyard or lever, and of the motion of the planets: that the weakening of power is in the ratio of the distances.

The sun is in the center of the planetary system.

But which body is it that is at the center? Is there none, as for Copernicus when he is computing, and for Tycho in part? Is it the earth, as for Ptolemy and for Tycho in part? Or finally, is it the sun itself, as I, and Copernicus when he is speculating, would have it? This question I began to discuss in physical terms in Part 1. I there supposed as one of the principles what has now been expressly and geometrically proved in Chapter 32: that a planet is moved less vigorously when it recedes from the point whence the eccentricity is computed.

From this principle I presented a probable argument that the sun is at that point and at the center of the world (or the earth for Ptolemy) rather than its being some other point occupied by no body. Allow me, then, to recall that same probable argument, its principles now demonstrated, in the present chapter. Then, as you may remember, I demonstrated in the second part, that the phenomena at either end of the night[1] follow beautifully if the oppositions of Mars are reckoned according to the sun's apparent position. If this is done, then we likewise set up the eccentricity and the distances from the very center of the sun's body, so that the sun itself again comes to be at the center of the world (for Copernicus), or at least at the center of the planetary system (for Tycho). But of these two arguments, one depends upon physical probability, and the other proceeds from possibility to actuality. And so in the third place I have demonstrated from the observations (in a proof which, because of its conceptual difficulty, I have postponed until Chapter 52) that we cannot avoid referring Mars to the apparent position of the sun, and drawing the line of apsides, which bisects the eccentric, directly through the sun's body, unless perhaps we wish to allow an eccentric such as will by no means be in accord with the parallax of the annual orb.[2] Anyone who cannot tolerate the delay may read about this in Chapter 52, and then may carry on here afterwards. For there nothing is assumed but the bare observations. You will find a similar proof in Part 5, from considerations of latitudes.

The motive power is in the sun.

Therefore, with the sun belonging in the center of the system, the source of motive power, from what has now been demonstrated, belongs in the sun, since it too has now been located in the center of the world.

But indeed, if this very thing which I have just demonstrated *a posteriori* (from the observations) by a rather long deduction, if, I say, I had taken this as something to be demonstrated *a priori* (from the worthiness and eminence of the sun), so that the source of the world's life (which is visible in the motion of the heavens) is the same as the

1. That is, when a planet is rising as the sun is setting, or *vice versa.*

2. Another name for the second inequality.

source of the light which forms the adornment of the entire machine, and which is also the source of the heat by which everything grows, I think I would deserve an equal hearing.

The sun is in the center of the world, and does not move from place to place.

Tycho Brahe himself, or anyone who prefers to follow his general hypothesis of the second inequality, should consider by how close a likeness to the truth this physically elegent combination has for the most part been accepted (since for him, too, this substitution of the apparent position of the sun brings the sun back to the center of the planetary system) yet to some extent recoils from his hypothesis.

For it is obvious from what has been said that only one of the following can be true: either the power residing in the sun, which moves all the planets, by the same action moves the earth as well; or the sun, together with the planets linked to it through its motive force, is borne about the earth by some power which is seated in the earth.

Now Tycho himself destroyed the notion of real orbs, and I in turn have in this third part irrefutably demonstrated that there is an equant in the theory of the sun or earth. From this it follows that the motion of the sun itself (if it is moved) is intensified and remitted according as it is nearer or farther from the earth, and hence that the sun is moved by the earth. But if, on the other hand, the earth is in motion, it too will be moved by the sun with greater or less velocity according as it is nearer or farther from it, while the power in the body of the sun remains perpetually constant. Between these two possibilities, therefore, there is no intermediate.

I myself agree with Copernicus, and allow that the earth is one of the planets.

Now it is true that the same objection may be raised against Copernicus concerning the moon, that I raised against Tycho concerning the five planets; namely, that it appears absurd for the moon to be moved by the earth, and to be associated with it and bound to it as well, so that it too, as a secondary planet, is swept around the sun by the sun. Nevertheless, I prefer to allow one moon, akin to the earth in its corporeal disposition (as I have shown in the *Optics*[3]) to be moved by a power seated in the earth but extended towards the sun, as will be described a little later in Chapter 37, rather than also to

3. *Optics* Chapter 6, pp. 242–3 in the Donahue translation.

ascribe to that same earth the motion of the sun and of all the planets bound to it.

The kinship of the solar motive power with light.

But let us carry on to a consideration of this motive power residing in the sun, and let us now again observe its very close kinship with light. Since the perimeters of similar regular figures, even of circles, are to one another as their semidiameters, therefore as *ad* is to *ae,* so is the circumference of the circle described about *a* through *d* to the circumference of the circle described about the same point *a* through *e.*[4] But as *ad* is to *ae,* so (inversely) is the strength of the power at *e* to the strength of the power at *d,* by what was proved in Chapter 32. Therefore, as the circle at *d* is to the smaller circle at *e,* so, inversely, is the power at *e* to the power at *d;* that is, the power is weaker to the extent that it is more spread out, and stronger to the extent that it is more concentrated. Hence we may understand that there is the same power in the whole circumference of the circle through *d* as there is in the circumference of the smaller circle through *e.* This is shown to be true of light in exactly the same way in the *Astronomiae pars optica,* Chapter 1.[5]

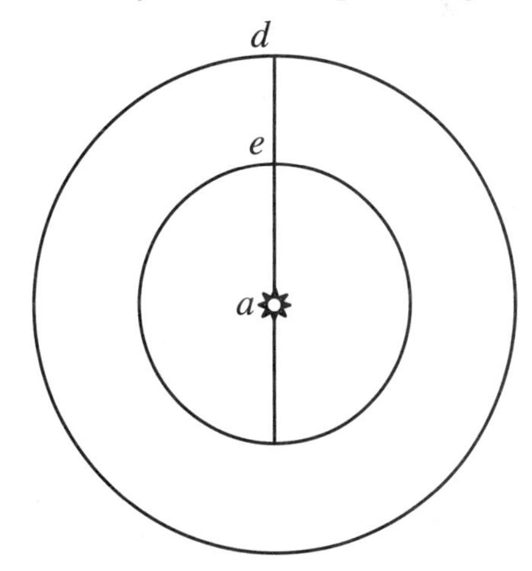

Therefore, in all respects and in all its attributes, the motive power from the sun coincides with light.

Whether light is the vehicle of the motive power.

And although this light of the sun cannot be the moving power itself, I leave it to others to see whether light may perhaps be so constituted as to be, as it were, a kind of instrument or vehicle, of which the moving power makes use.

4. For reasons of economy, Kepler reused a complicated diagram from Chapter 32 here. The present diagram is simplified to fit his proof.

5. Kepler, *Optics,* Chapter 1 Proposition 9, Donahue trans. p. 22.

This seems gainsaid by the following: first, light is hindered by the opaque, and therefore if the moving power had light as a vehicle, darkness would result in the movable bodies being at rest; again, light spreads spherically in straight lines, while the moving power, though spreading in straight lines, does so circularly; that is, it is exerted in but one region of the world, from east to west, and not the opposite, not at the poles, and so on. But we shall be able to reply plausibly to these objections in the chapters immediately following.

The moving power is an immaterial *species*[6] of the solar body.

Finally, since there is just as much power in a larger and more distant circle as there is in a smaller and closer one, nothing of this power is lost in travelling from its source, nothing is scattered between the source and the movable body. The emission, then, in the same manner as light, is immaterial, unlike odors, which are accompanied by a diminution of substance, and unlike heat from a hot furnace, or anything similar which fills the intervening space. The remaining possibility, then, is that, just as light, which lights the whole earth, is an immaterial *species* of that fire which is in the body of the sun, so this power which enfolds and bears the bodies of the planets, is an immaterial *species* residing in the sun itself, which is of inestimable strength, seeing that it is the primary agent of every motion in the universe.

Since, therefore, this *species* of the power, exactly as the *species* of light (for which see the *Astronomiae pars optica* Chapter 1),[7] cannot be considered as dispersed throughout the intermediate space between the source and the mobile body, but is seen as collected in the body in proportion to the amount of the circumference it occupies, this power (or *species*) will therefore not be any geometrical [solid] body, but is like a variety of surface, just as light is. To generalize this, the *species* which proceed immaterially from things are not by that procession extended through the dimensions of a body, although they arise from a body (as this one does from the body of the sun). Instead, they proceed according to that very law of emission: they do not possess their own boundaries, but just as the surfaces of illuminated things cause light to be considered as surfaces in certain respects, because they receive and

6. This word has such a range of meanings that it seems best to leave it untranslated. See the Glossary for a more complete account.

7. *Optics* Chapter 1, Donahue trans. p. 19.

terminate its emission, so the bodies of things that are moved suggest that this moving power be considered as if a sort of geometrical [solid] body, because their whole masses terminate or receive this emission of the motive *species,* so that the *species* can exist or subsist nowhere in the world but in the bodies of the mobile things themselves. And, exactly like light, between the source and the movable thing it is in a state of becoming, rather than of being.

In what manner the immaterial *species* of the body of the sun may be quantified.

Moreover, at the same time, a reply can be made here to a possible objection. For it was said above that this motive power is extended throughout the space of the world, in some places more spread out and in others more concentrated, and that the intensification and remission of the motions of the planets are consequent upon this variation. Now, however, it has been said that this power is an immaterial *species* of its source, and never inheres in anything except a mobile subject, such as the body of a planet. But to lack matter and yet to be subject to geometrical dimensions appear to be contradictory. This implies that it is poured out throughout the whole world, and yet does not exist anywhere but where there is something movable.

The reply is this: although the motive power is not anything material, nevertheless, since it is destined to carry matter (namely, the body of a planet), it is not free from geometrical laws, at least on account of this material action of carrying things about. Nor is there need for more, for we see that those motions are carried out in space and time, and that this power arises and is poured out from the source through the space of the world, all of which are geometrical entities. So this power should indeed be subject to other geometrical necessities.

Comparison of light and motive power on the basis of quantification and of time

But lest I appear to philosophize with excessive insolence, I shall propose to the reader the clearly authentic example of light, since it also makes its nest in the sun, thence to break forth into the whole world as a companion to this motive power. Who, I ask, will say that light is something material? Nevertheless, it carries out its operations with respect to place, suffers alteration, is reflected and refracted, and assumes quantities so as to be dense or rare, and to be capable of being taken as a surface wherever it falls upon something illuminable. Now just as it is said in optics, that light does not exist in the intermediate space between the source and the illuminable, this is equally true of the motive power. Moreover, although light itself does indeed flow forth in no time, while this power creates motion in time, nonetheless the way in which both do so is the same, if you consider them correctly. Light manifests those things which are proper to it instantaneously, but requires time to effect those which are associated with matter. It illuminates a surface in a moment, because here matter need not undergo any alteration, for all illumination takes place according to surfaces, or at least as if a property of surfaces and not as a property of corporeality as such. On the other hand, light bleaches colors in time, since here it acts upon matter *qua* matter, making it hot and expelling the contrary cold which is embedded in the body's matter and is not on its surface. In precisely the same manner, this moving power perpetually and without any interval of time is present from the sun wherever there is a suitable movable body, for it receives nothing from the movable body to cause it to be there. On the other hand, it causes motion in time, since the movable body is material.

Why the planets do not equal their mover, the immaterial *species* of the sun, in speed.

Or if it seems better, frame the comparison in this manner: light is constituted for illumination, and it is just as certain that power is constituted for motion. Light does everything that can be done to achieve the greatest illumination; nonetheless, it does not happen that color is most greatly illuminated. For color intermingles its own peculiar *species* with the illumination of light, thus forming some third entity. In like manner, there is no retardation in the moving power to prevent the planet's having as much speed as it has itself, but it does not follow that the planet's speed is that great, since something intervening prevents that, namely, some sort of matter possessed by the surrounding aether, or the disposition of the movable body itself to rest (others might say, "weight," but I do not entirely approve of that, except, indeed, where the earth is concerned). It is the tempering effect of these, together with the weakening of the motive power, that determines a planet's periodic time.

Chapter 34
The sun is a magnetic body and rotates in its space

Concerning that power that is closely attached to, and draws, the bodies of the planets, we have already said how it is formed, how it is akin to light, and what it is in its metaphysical being. Next, we shall contemplate the deeper nature of its source, shown by the outflowing *species* (or archetype). For it may appear that there lies hidden in the body of the sun a sort of divinity, which may be compared to our soul, from which flows that *species* driving the planets around, just as from the soul of someone throwing pebbles a *species* of motion comes to inhere in the pebbles thrown by him, even when he who threw them removes his hand from them. And to those who proceed soberly, other reflections will soon be provided.

The power that moves the planets is whirled around.

The power that is extended from the sun to the planets moves them in a circular course around the immovable body of the sun. This cannot happen or be conceived in thought in any other way than this, that the power traverses the same path along which it carries the other planets. This has been observed to some extent in catapults and other violent motions. Thus, Fracastoro and others, relying on a story told by the most ancient Egyptians, spoke with little probability when they said that some of the planets perchance would have their orbits deflected gradually beyond the poles of the world, and thus afterwards would move in a path opposite to the rest and to their modern course.[1] For it is much more likely that the bodies of the planets are always borne in that direction in which the power emanating from the sun tends.

But this *species* is immaterial, proceeding from its body out to this distance without the passing of any time, and is in all other respects like light. Therefore, it is not only required by the nature of the *species,* but likely in itself owing to this kinship with light, that along with the particles of its body or source it too is divided up, and when any particle of the solar body moves towards some part of the world, the particle of the immaterial *species* that from the beginning of creation corresponded to that particle of the body also always moves towards

1. Hieronymus Fracastorius, *Homocentrica,* Venice 1538, Sect. 3 Cap. 8.

the same part. If this were not so, it would not be a *species*, and would come down from the body in curved rather than straight lines.

Since the *species* is moved in a circular course, in order thereby to confer motion upon the planets, the body of the sun, or source, must move with it, not, of course, from space to space in the world—for I have said, with Copernicus, that the body of the sun remains in the center of the world—but upon its center or axis, both immobile, its parts moving from place to place, while the whole body remains in the same place.

Example from light.

In order that the force of the analogical argument may be that much more evident, I would like you to recall, reader, the demonstration in optics that vision occurs through the emanation of small sparks of light[2] toward the eye from the surfaces of the seen object. Now imagine that some orator in a great crowd of people, encircling him in an orb,[3] turns his face, or his whole body along with it, once around. Those of the audience to whom he turns his eyes directly will also see his eyes, but those who stand behind him then lack the view of his eyes. But when he turns himself around, he turns his eyes around to everyone in the orb. Therefore, in a very short interval of time, all get a glimpse of his eyes. This they get by the arrival of a spark of light or *species* of color descending from the eyes of the orator to the eyes of the spectators. Thus by turning his eyes around in the small space in which his head is located, he carries around along with it the rays of

---

2. *Luculae.* Kepler uses this odd word in his *Optics* Chapter 1 Propositions 15 and 32 (Donahue trans. pp. 24 and 40). He says that in total darkness colored objects may still emit *luculae*, and that there probably is a *lucula* in the heart accompanying the heart's fire. Evidently, a *lucula* is a very small spark of light. It is not a "ray," for a ray "is nothing but the motion of light" (*Optics* Chapter 1 Prop. 8, Donahue trans. p. 22), nor is it a particle in the usual sense of the word, since it "lacks corporeal matter, but consists of its own sort of matter" (*Optics* Donahue trans. p. 19). The chief distinguishing feature of this matter is that it is "a kind of surface" (*Optics* Chapter 1 Prop. 8, Donahue trans. p. 22). So it might be most nearly correct to imagine a *lucula* as a two dimensional "particle" or very small part of the lucid *species* of a body, whose motion (at right angles to its surface) constitutes a "ray."

3. In Latin, as in English, the word *orbis* (orb) is ambiguous: it can denote either a circle or a sphere. Care has therefore been taken to preserve the ambiguity in the translation.

the spark of light in the very large orb in which the eyes of the spectators all around are situated. For unless the spark of light went around, his spectators would not be recipients of his eyes' glance. Here you see clearly that the immaterial *species* of light either is moved around or stands still depending upon whether that of which it is the *species* either moves or stands still.[4]

The sun rotates.

Therefore, since the *species* of the source, or the power moving the planets, rotates about the center of the world, I conclude with good reason, following this example, that that of which it is the *species,* the sun, also rotates.[5]

However, the same thing is also shown by the following argument. Motion that is local and subjected to time cannot inhere in a bare immaterial *species,* since such a *species* is incapable of receiving an applied motion unless the received motion is nontemporal, just as the power is immaterial. Also, although it has been proved that this moving power rotates, it cannot be allowed to have infinite speed (for then it would seem that infinite speed would also have to be imposed upon the bodies), and therefore it completes its rotation in some period of time. Therefore, it cannot carry out this motion by itself, and it is as a consequence necessary that it is moved only because the body upon which it depends is moved.

By the same argument, it appears to be a correct conclusion that there does not exist within the boundaries of the solar body anything immaterial by whose rotation the *species* descending from that immaterial something also rotates. For again, local motion which takes time cannot correctly be attributed to anything immaterial. It therefore remains that the body of the sun itself rotates in the manner described above, indicating the poles of the zodiac by the poles of its rotation

4. In the *Dialogue on the Two Chief World Systems,* Galileo has Salviati say the following: "So the turning motion made by the fowling piece in following the flight of the bird with the sights, though slow, must be communicated to the ball also;..." (p. 178 in the Stillman Drake translation). Evidently, the notion of a sort of circular impetus or inertia had its attractions. However, Galileo's interlocutors reject the idea shortly after.

5. At the time of publication, there was no observational evidence that the sun rotates: Galileo first turned the telescope on the heavens a year later. Thus when Kepler later heard of Galileo's discovery, he regarded it as a dramatic confirmation of his celestial physics.

(by extension to the fixed stars of the line from the center of the body through the poles), and indicating the ecliptic by the greatest circle of its body, thus furnishing a natural cause for these astronomical entities.

In the bodies of the planets is a material inclination to rest in every place where they are put by themselves.

Further, we see that the individual planets are not carried along with equal swiftness at every distance from the sun, nor is the speed of all of them at their various distances equal. For Saturn takes 30 years, Jupiter 12, Mars 23 months, earth 12, Venus eight and one half, and Mercury three. Nevertheless, it follows from what has been said that every orb of power emanating from the sun (in the space embraced by the lowest, Mercury, as well as that embraced by the highest, Saturn) is twisted around with a whirl equal to that which spins the solar body, with an equal period. (There is nothing absurd in this statement, for the emanating power is immaterial, and by its own nature would be capable of infinite speed if it were possible to impress a motion upon it from elsewhere, for then it could be impeded neither by weight, which it lacks, nor by the obstruction of the corporeal medium.) It is consequently clear that the planets are not so constituted as to emulate the swiftness of the motive power. For Saturn is less receptive than Jupiter, since its returns are slower, while the orb of power at the path of Saturn returns with the same swiftness as the orb of power at the path of Jupiter, and so on in order, all the way to Mercury, which, by example of the superior planets, doubtless moves more slowly than the power that pulls it. It is therefore necessary that the nature of the planetary globes be material, from an inherent property, arising from the origin of things, to be inclined to rest or to the privation of motion. When the tension between these things leads to a fight, the power [from the sun] is more overcome by that planet which is placed in a weaker power, and is moved more slowly by it, and is less overcome by a planet that is closer to the sun.

This analogy shows that there is in all planets, even in the lowest, Mercury, a material force of disengaging itself somewhat from the orb of the sun's power.

The amount of time in which the rotation of the solar body traverses its space.

From this it is concluded that the rotation of the solar body anticipates considerably the periodic times of all the planets; therefore, it must rotate in its space at least once in a third of a year.

However, in my *Mysterium Cosmographicum* I pointed out that there is about the same ratio between the semidiameters of the sun's body and the orb of Mercury as there is between the semidiameters of the body of the earth and the orb of the moon. Hence, you may plausibly conclude that the period of the orb of Mercury would have the same ratio to the period of the body of the sun as the period of the orb of the moon has to the period of the body of the earth. And the semidiameter of the orb of the moon is sixty times the semidiameter of the body of the earth, while the period of the orb of the moon (or the month) is a little less than thirty times the period of the body of the earth (or day), and thus the ratio of the distances is double the ratio of the periodic times. Therefore, if the doubled ratio also holds for the sun and Mercury, since the diameter of the sun's body is about one sixtieth of the diameter of Mercury's orb, the time of rotation of the solar globe will be one thirtieth of 88 days, which is the period of Mercury's orb. Hence it is likely that the sun rotates in about three days.

Whether the earth's diurnal rotation comes from the rotation of the solar globe.

You may, on the other hand, prefer to prescribe the sun's diurnal period in such a way that the diurnal rotation of the earth is dispensed by the diurnal rotation of the sun, by some sort of magnetic force. I would certainly not object. Such a rapid rotation appears not to be alien to that body in which lies the first impulse for all motion.

The monthly motion of the moon arises from the diurnal rotation of the earth.

This opinion (on the rotation of the solar body as the cause of the motion for the other planets) is beautifully confirmed by the example of the earth and the moon. For the chief, monthly motion of the moon, by the force of the demonstrations used in Chapters 32 and 33, takes its origin entirely from the earth (for what the sun is for the rest of the planets there, the earth is for the moon in this demonstration). Consider, therefore, how our earth occasions the motion of the moon: while this our earth, and its immaterial *species* along with it, rotates twenty-nine and one-half times about its axis, at the moon this *species* has the capability of driving it only once around in the same time, in (of course) the same direction in which the earth leads it.

Here, by the way, is a marvel: in any given time the center of the moon traverses twice as long a line about the center of the earth as any place on the surface of the earth beneath the great circle of the equator. For if equal spaces were measured out in equal times, the moon ought to return in sixty days, since the size of its orb is sixty times the size of the earth's globe.[6]

This is surely because there is so much force in the immaterial *species* of the earth, while the lunar body is doubtless of great rarity and weak resistance. Thus, to remove your bewilderment, consider that on the principles we have supposed it would necessarily follow that if the moon's material force had no resistance to the motion impressed from outside by the earth, the moon would be carried at exactly the same speed as the earth's immaterial *species*, that is, with the earth itself, and would complete its circuit in 24 hours, in which the earth also completes its circuit. For even if the tenuity of the earth's *species* is great at the distance of 60 semidiameters, the ratio of one to nothing is still the same as the ratio of sixty to nothing.[7] Hence the immaterial *species* of the earth would win out completely, if the moon did not resist.

What sort of body is the sun?

Here, one might inquire of me, what sort of body I consider the sun to be, from which this motive *species* descends. I would ask him to proceed under the guidance of a further analogy, and urge him to inspect more closely the example of the magnet brought up a little earlier, whose power resides in the entire body of the magnet when it grows in mass, or when by being divided it is diminished. So in the sun the moving power appears so much stronger that it seems likely that its body is of all [those in the world] the most dense.

---

6. A point on the earth's equator takes one day to complete one revolution. Astronomers were in agreement that the moon's distance from the center of the earth is about 60 times the earth's radius; therefore, in one revolution, the moon travels about 60 times as far as the point on the earth's equator travels in one revolution. If the moon and the point on earth were going at the same speed, the moon would therefore take 60 days to complete its orbit. But the moon takes only about 30 days to get around; therefore, it is going about twice as fast as the point on the earth's equator.

7. Kepler's idea is that the speed of a body is proportional to ratio of the moving power to the body's resistance to motion. If the resistance is zero, the ratio is the same (i. e., infinite), regardless of the power at different distances.

Likeness of the sun's body to a magnet.

And the power of attracting iron is spread out in an orb from the magnet so that there exists a certain orb within which iron is attracted, but more strongly so as the iron comes nearer into the embrace of that orb. In exactly the same way the power moving the planets is propagated from the sun in an orb, and is weaker in the more remote parts of the orb.

The difference between the solar body and a magnet.

The magnet, however, does not attract with all its parts, but has filaments (so to speak) or straight fibers (seat of the motor power) extended throughout its length, so that if a little strip of iron is placed in a middle position between the heads of the magnet at the side, the magnet does not attract it, but only directs it parallel to its own fibers. Thus it is credible that there is in the sun no force whatever attracting the planets, as there is in the magnet, (for then they would approach the sun until they were quite joined with it), but only a directing force, and consequently that it has circular fibers all set up in the same direction, which are indicated by the zodiac circle.

The principle of motion in the sun and in the magnet is the same.

Therefore, as the sun forever turns itself, the motive force or the outflowing of the *species* from the sun's magnetic fibers, diffused through all the distances of the planets, also rotates in an orb, and does so in the same time as the sun, just as when a magnet is moved about, the magnetic power is also moved, and the iron along with it, following the magnetic force.

By example of the earth, it is proved that there are magnets in the heavens.

The example of the magnet I have hit upon is a very pretty one, and entirely suited to the subject; indeed, it is little short of being the very truth. So why should I speak of the magnet as if it were an example? For, by the demonstration of the Englishman William Gilbert,[8] the earth itself is a big magnet, and is said by the same

---

8. William Gilbert, *De magnete magneticisque corporibus et de magno Magnete Tellure physiologia nova,* London, 1600. English translation by P. Fleury Mottelay (New York: Wiley, 1893, repr. Dover, 1958 etc.).

author, a defender of Copernicus, to rotate once a day, just as I conjecture about the sun. And because of that rotation, and because it has magnetic fibers intersecting the line of its motion at right angles, those fibers lie in various circles about the poles of the earth parallel to its motion. I am therefore absolutely within my rights to state that the moon is carried along by the rotation of the earth and the motion of its magnetic power, only thirty times slower.

Likeness of the earth and the sun, with respect to the motion impressed upon the planets.

I know that the earth's filaments and its motion indicate the equator, while the circuit of the moon is generally related to the zodiac—on this point there will be more in Chapter 37 and Part 5. With this one exception, everything fits: the earth is intimately related to the lunar period, just as the sun is to that of the other planets. And just as the planets are eccentric with respect to the sun, so is the moon with respect to the earth. So it is certain that the earth is looked upon by the moon's mover as its pole star (so to speak), just as the sun is looked upon by the movers belonging to the rest of the planets, for which see Chapter 38. It is therefore plausible, since the earth moves the moon through its *species* and is a magnetic body, while the sun moves the planets similarly through an emitted *species*, that the sun is likewise a magnetic body.

Chapter 39
By what path and by what means do the powers seated in the planets need to move them in order to produce a planetary orbit through the aethereal air that is circular, as it is commonly thought to be

And so, in what has been demonstrated, let us take these axioms, which are of great certainty. First, that the body of a planet is inclined by nature to rest in every place where it is put by itself. Second, that it is transported from one longitudinal position to another by that power that originates in the sun. Third, if the distance of the planet from the sun were not altered, a circular path would result from this motion. Fourth, supposing the same planet to be in turn at two distances from the sun, remaining there for one whole circuit, the periodic times will be in the duplicate ratio of the distances or magnitudes of the circle.[1] Fifth, the bare and solitary power residing in the body of a planet itself is not sufficient for transporting its body from place to place, since it lacks feet, wings, and feathers by which it might press upon the aethereal air. And nevertheless, sixth, the approach and recession of a planet to and from the sun arises from that power that is proper to the planet. All these axioms are agreeable to nature in themselves, and have been demonstrated previously.

Now let us work with geometrical figures in order to see what laws will be required to represent any desired planetary orbit. Let the orbit of the planet be a circle, as has been believed until now, and let it be eccentric with respect to the sun, the source of power.

At this point, Kepler describes three ways that a planet's mover might trace out a circular orbit. All of them involve the planet moving on a small circle or epicycle that is swung around the sun. The epicycle serves to adjust the planet's distance, while the sun's power carries it around. The first two options require real, physical epicycle. However, recalling what he had said in Chapter 2, he rejected real epicycles as absurd. His third alternative, then, was that the planetary mover or moving power might somehow follow the pattern of an *imaginary* epicycle. It is as if

1. This is not proved explicitly, but is a consequence of the demonstrations in Chapters 32 and 33. A demonstration is provided in Appendix B, p. 101.

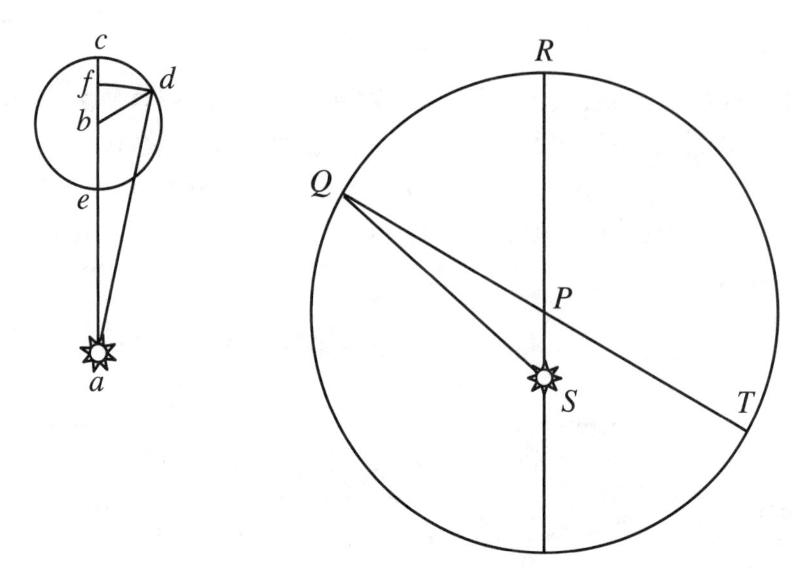

the planet, while it moves around the sun, reciprocates in and out along the line *ac* in such a way that its distance from the sun is represented by *ad* or *af.*

Here, as in many places in *Astronomia Nova,* Kepler shifts, perhaps confusingly, between this epicyclic model and the geometrically equivalent eccentric model. In the eccentric model, the sun is at *S,* and the planet moves around the circumference of eccentric circle *RQT,* whose center is *P.* If *PR* and *PQ* are made equal to *ab* in the epicyclic model, and *PS* is equal to *bd,* while the angle of eccentric anomaly *QPR* or *SPT* is equal to angle *cbd,* then the two models are geometrically equivalent, and distance *QS* is always equal to distance *ad.* Kepler accordingly will sometimes refer to the angle *dbc* or the arc *dc* on the epicycle as the "eccentric anomaly," because it is numerically equal to the eccentric anomaly *RPQ.*

We pick up Kepler's account as he critiques the idea of reciprocation on an imaginary epicycle.

What, then, if we should say this: although the motion of the planet is not physically on an epicycle, this reciprocation is measured out in such a way that distances from the sun are arrived at which are similar to those which exist in an epicycle actually traversed?

First, this would be to attribute to the power belonging to the planet a knowledge of the imaginary epicycle and of its effects in setting out distances from the sun; and further, it would attribute knowledge of the future speed or slowness which the common motion from the sun is going to cause. For it is necessary to suppose here an imaginary intensification and remission of motion on the imaginary epicycle that is the same as that of the motion on the real eccentric. This is more incredible than the previous accounts, where the motion of the body was combined with knowledge of the epicycle or eccentric. Therefore, the objections raised there should be understood as applying here as well, the judgments being nearly identical.

Nevertheless, for want of a better opinion, we must at present put up with this one. And as for its involving many absurdities, a certain physicist[2] will allow, in Chapter 57 below, that on the testimony of the observations the path of the planet is not a circle.

By what means or measure may a planet grasp its distance from the sun?

So far, the discussion has concerned the measure relating to the form of this reciprocation. It now remains for us to find the measure of this measure; that is, the quantity of local motion. For it is not enough for the planet to know how far it should be from the sun: it also has to know what to do in order to be at the correct distance.

Now anyone who is so attracted to the supposition of a perfectly circular orbit as to associate a mind with the planet which could preside over the reciprocation, can say only this: that this planetary mind observes the increasing and decreasing size of the solar diameter, and understands, using this as an indication, what distances from the sun it should arrive at at any given time. For example, sailors cannot know from the sea itself how far they have travelled over the waters, since the course, viewed in that way, has no distinct limits. Instead, they find this either from the amount of time they have sailed, if wind and sea remain constant and the ship does not stop, or from the direction of the wind and the changing elevation of the pole, or from all or several of these in conjunction, or—honest to God—by a contrivance of a number of wheels, with paddles lowered into the water (for certain conceited mechanics are proposing an instrument of this sort, who ascribe the calm of the continents to the water of the Ocean). In just the same way, the mind of the planet cannot by itself measure its position, or the distance between itself and the sun, since between them there is pure aethereal air, devoid of any means of indication. So it must either make use of the elapsed time, in conjunction with a supposed invariance of forces (which has just been denied above),

2. Kepler himself.

or of a physical machine, which is ridiculous (for by the example of the sun and moon we suppose the celestial bodies to be round, and it is therefore also probable that the entire field of the aetheral air moves around with the planets),[3] or finally, of some suitable means of indication that varies with the distance of the planet from the sun. And other than the single indication of the sun's apparent diameter, nothing else presents itself. Thus we humans know that the sun is 229 of its own semidiameters distant from us when its diameter subtends 30', and 222 semidiameters when it subtends 31'.[4]

If it were indeed certain that this motion of the planet along the diameter of the epicycle could not be carried out by any material and corporeal or magnetic power of the planet, nor by an unassisted animate power, but that it is governed by a planetary mind,[5] nothing absurd would be stated. For that the sun is observed by the planets in other respects as well, the motions in latitude bear witness. For by these motions the planets would depart from the middle and royal road of this solar power, as from the mainstream of a river, and move to the sides, as is said in Chapter 38, unless they meanwhile paid attention to the sun, approaching and receding along a line drawn to its center. It would then describe circles which, seen from the earth or the center of the world, would appear smaller, parallel to some great circle. But all planets describe great circles which intersect the ecliptic at points that are opposite with respect to the sun, as was demonstrated for Mars from observations above in Chapters

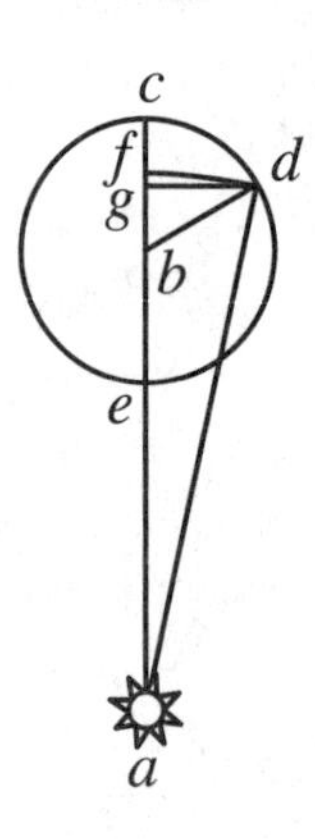

3. That is, it would be of no use to know the relative velocity of a planet through the aethereal air.

4. There's no need to understand how Kepler gets these particular numbers: his point here is that if a human observer can figure these things out, then an intelligent mover could too. (If you still want to know where the numbers come from, an explanation is given in Appendix B, p. 102.)

5. This was the standard view of the day, included in all the introductory university textbooks in natural philosophy. The planetary mind was usually identified with the biblical angels, thus bringing Aristotle into harmony with scripture. However, some theorists believed that the mind is united with the planet as soul is with body. This view was widely regarded as heretical, as it suggested that the Prime Mover is the soul of the world (*anima mundi*), and thus that the universe is somehow God's body.

12, 13, and 14. Therefore, the diameter of reciprocation *ce* is also directed towards the sun, and the latitudes are related to the sun in every respect. This remains true despite my ascribing the latitudinal motions partly to mind and partly to nature and magnetic faculties, in Part 5 below.

Now one cannot say in reply to me that the solar diameter and its variation are far too small to be used as a standard. For it is certain that there is no planet for which it entirely vanishes. Since on earth it is thirty minutes, on Mars it will be twenty, on Jupiter seven, and on Saturn three, while on Venus it will be forty, and on Mercury at least eighty and sometimes as high as one hundred and twenty. The query should not concern the smallness of the body, but rather the unapt coarseness of human perception, which cannot be stretched to sense such small things.

One should on the contrary note that this body, however small, is nonetheless capable of moving such distant bodies in a circle, as is demonstrated in the preceding chapters. The illumination of the world by such a tiny corpuscle is known to all. And so it is credible that if the movers are endowed with some faculty of observing its diameter, this faculty is as much more acute than our eyes as its work, and the perpetual motion, is more constant than our own troubled and confused schemes.

So then, Kepler, would you give each of the planets a pair of eyes? By no means, nor is this necessary, no more than that they need feet or wings in order to move. But Brahe has recently eliminated solid orbs. Now our theorizing has not emptied nature's treasure house: we still cannot establish, through our own knowledge, how many senses there ought to be. There are even examples at hand worthy of our admiration. For tell us, in physical terms: with what eyes shall the animate faculties of sublunar bodies look upon the positions of the stars in the zodiac, so that when a harmonic arrangement (which we call "aspects") is found among them, the bodies leap up and display it in their actions? Was it with her eyes that my mother noted the positions of the stars in order to know that she was born with Saturn, Jupiter, Mars, Venus, and Mercury, all in sextiles and trines? And could it have been by the same means that her children, and especially I her first born, came to see the light chiefly on those days when as many as possible of the same aspects, especially of Saturn and Jupiter, recurred, or when they possessed as many pristine positions as possible, with squares, oppositions, and conjunctions? I have observed those things in all cases whatever that have occurred to this very day. But what is to be made of these examples, just as absurd as the previous ones? The answer will have to await those who work harder in their study of nature than is usual today.

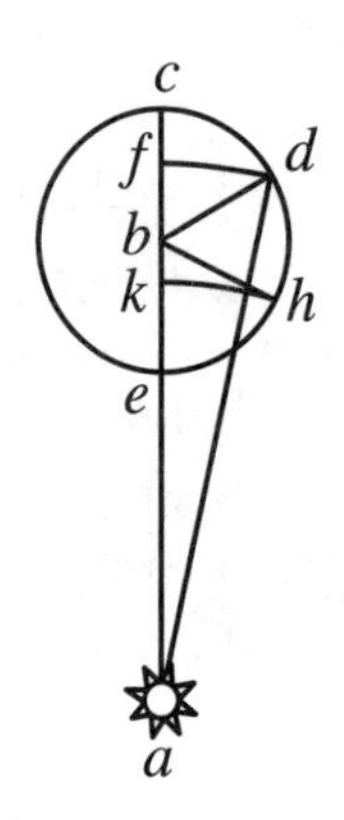

So the hypothetical person who says that the planet's path is a perfect circle will say this: that the planet performs its reciprocation so as to make the diameter of the sun, at the end points of equal arcs of the eccentric, appear very nearly inversely proportional to the lines *da, ha, ea,* or to *fa, ka, ea* which are equal to them, taken with respect to the longest line *ca;* and that through this consideration of the diameter of the sun at the chosen moments of time, come the distances of *f, k, e* from *c.*

It should be made known, however, that the increases of the diameter of the sun and the arcs of the epicycle do not square with each other well, and so the motive mind will have to have a very good memory in order to adjust the unequal versed sines[6] of the arcs on the epicycle to the equal increases of the solar diameter. For this, see Chapters 56 and 57 below.

Let that be enough concerning the means of indicating distances. There remains a third topic to which I wish briefly to draw attention: the nature of the animate faculty that carries the planet about. Anyone who says that the body of a planet is moved by an inherent force is just plain wrong. This we proved at the beginning. But it is likewise impossible simply to ascribe this force to the sun instead. For the same force that attracts the planet also repels it in turn, and this is inconsistent with the simplicity of the solar body. But anyone who by some unique argument reduces this motion to the bodies of the sun and the planet in concerted action gives a new cast to the material

6. The "versed sines" here are the lines *cf, ck, ce* that measure how far the planet descends from aphelion towards the sun. If the arcs *cd, dh, he* are made equal, the corresponding increments in the versed sines *cf, fk, ke* are clearly not equal. Furthermore, as Kepler was well aware, there is no direct way to calculate these increments from the arcs: approximations have to be used, and the results set out in tables. This is why Kepler says that the motive mind would have to have a very good memory: it would have to know the tables by rote.

For those who wish to know more of the mathematical details of this remark, there is a note in Appendix B, p. 102.

of this entire chapter, and is consequently referred to this particular topic in Chapter 57 below.

You see, my thoughtful and intelligent reader, that the opinion of a perfect eccentric circle for the path of a planet drags many incredible things into physical theories. This is not, indeed, because it makes the solar diameter an indicator for the planetary mind, for this opinion will perhaps turn out to be closest to the truth, but because it ascribes incredible faculties to the mover, both mental and animate.

Although our theories are not yet complete and perfect, they are nearly so, and in particular are suitable for the motion of the sun,[7] so we shall pass on to quantitative consideration. And when at last we approach a more exact discovery of the truth, reserved for Chapter 57, it will be seen that we laid the groundwork for it here.

7. That is, the construction of the earth's orbit, which is more nearly circular than Mars's orbit is.

Chapter 40
An imperfect method for computing the equations from the physical hypothesis, which nonetheless suffices for the theory of the sun or earth.

In this chapter, Kepler further develops the principle he established in Chapter 32. There he had argued that Ptolemy's equant is really a flawed geometrical expression of the dynamic principle that planets move faster as they get nearer the sun. The question in the present chapter is how to incorporate this qualitative speed rule into a mathematical model that can give give accurate planetary positions. The result, presented here, is the computational rule that we now know as "Kepler's Second Law": that radii drawn from planets to the sun sweep out areas proportional to the times.

Such a long-winded discussion was necessary to prepare a way for a natural form for the equations, on which I am going to be very busy in Part 4 [Chapters 41–60]. Now we must return to the equations of the sun's eccentric in particular, which is the main subject of this third part, and for the sake of which the general discussion of the last eight chapters has been presented.

My first error was to suppose that the path of the planet is a perfect circle, a supposition that was all the more noxious a thief of time the more it was endowed with the authority of all philosophers, and the more convenient it was for metaphysics in particular. Accordingly, let the path of the planet be a perfect eccentric [circle], for in the theory of the sun the amount by which it differs from the oval path is imperceptible. Those things that are going to be needed for the other planets, on account of this deviation, follow below in Chapters 59 and 60.

Since, therefore, the times of a planet over equal parts of the eccentric are to one another as the distances of those parts, and since the individual points of the entire semicircle of the eccentric are all at different distances, it was no easy task I set myself when I sought to find how the sums of the individual distances may be obtained. For unless we can find the sum of all of them (and they are infinite in number) we cannot say how much time has elapsed for any one of them. Thus the equation will not be known. For the whole sum of the distances is to the whole periodic time as any partial sum of the distances is to its corresponding time.

I consequently began by dividing the eccentric into 360 parts, as if these were least particles, and supposed that within one such part the distance does not change. I then found the distances at the beginnings of the parts or degrees by the method of Chapter 29, and added them all up. Next, I assigned an artificial round number to the periodic time: although it is in fact 365 days and 6 hours, I set it equal to 360 degrees, or a full circle, which for the astronomers is the mean anomaly.[1] As a result, I have so arranged it that as the sum of the distances is to the sum of the time, so is any given distance to its time. Finally, I added the times over the individual degrees and compared these times, or degrees of mean anomaly, with the degrees of the eccentric anomaly, or the number of parts whose distance was sought. This furnished the physical equation, to which the optical equation, found by the method of Chapter 29 with those same distances, was to be added in order to have the whole.

However, since this procedure is mechanical and tedious, and since it is impossible to compute the equation given the ratio for one individual degree without the others, I looked around for other means. And since I knew that the points of the eccentric are infinite [in number], and their distances are infinite, it struck me that all these distances are contained in the plane of the eccentric. For I had remembered that Archimedes, in seeking the ratio of the circumference to the diameter, once thus divided a circle into an infinity of triangles—this being the hidden force of his *reductio ad absurdum*[2]. Accordingly, instead of dividing the circumference as before, I now cut the plane of the eccentric into 360 parts by lines drawn from the point whence the eccentricity is reckoned.

Let *AB* be the line of apsides, *A* the sun (or earth, for Ptolemy); *B* the center of the eccentric *CD*, whose semicircle *CD* shall be divided into any number of equal parts *CG, GH, HE, EI, IK, KD,* and let the points *A* and *B* be connected with the points of division. Therefore, *AC* will be the greatest distance, *AD* the least, and the others, in order,

1. For "mean anomaly," "eccentric anomaly," "equated anomaly," and other technical terms, see the Glossary and Appendix A.

2. Archimedes, *Measurement of a Circle* Prop. 1, in *The Works of Archimedes, with the Method of Archimedes,* ed. T. L. Heath (Cambridge 1912 and other eds.), pp. 91–93.

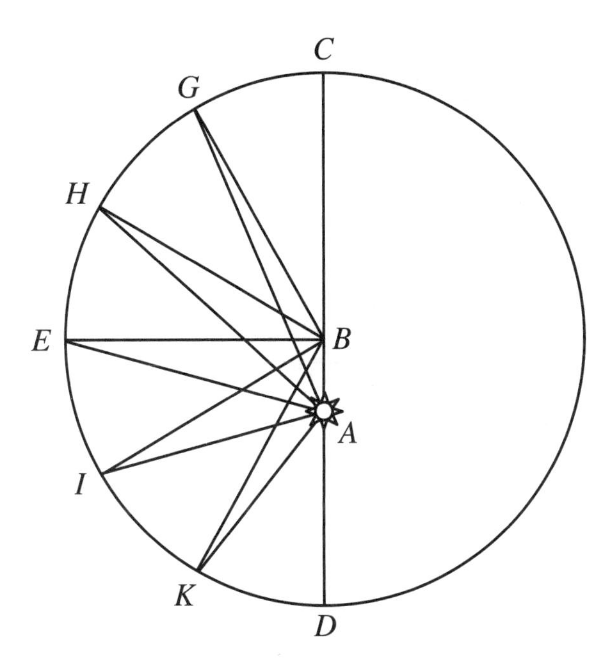

are *AG, AH, AE, AI, AK*. And since triangles under equal altitudes are as their bases, and the sectors, or triangles, *CBG, GBH,* and so on (standing upon least parts of the circumference and therefore not differing from straight lines) all have the same altitude, the equal sides *BC, BG, BH,* they are therefore all equal. But all the triangles are contained in the area *CDE,* and all the arcs or bases are contained in the circumference *CED*. Therefore, by composition, as the area *CDE* is to the arc *CED* so is the area *CBG* to the arc *CG,* and alternately, as arc *CED* is to *CG, CH,* and the rest in order, so is the area *CDE* to the areas *CBG, CBH,* and the rest in order. Therefore, no error is introduced if the areas be taken for the arcs in this way, and substituting the areas *CGB, CHB* for the angles of eccentric anomaly *CBG, CBH*.

Further, just as the straight lines from *B* to the infinite parts of the circumference are all contained in the area of the semicircle *CDE,* and the straight lines from *B* to the infinite parts of the arc *CH* are all contained in the area *CBH,* therefore also the straight lines from *A* to the same infinite parts of the circumference or arc make up the same thing. And finally, since those drawn from *A* and *B* both fill up one and the same semicircle *CDE,* while those from *A* are the very distances whose sum is sought, it therefore seemed to me I could conclude that by computing the area *CAH* or *CAE* I would have the sum of the infinite distances in *CH* or *CE,* not because the infinite can

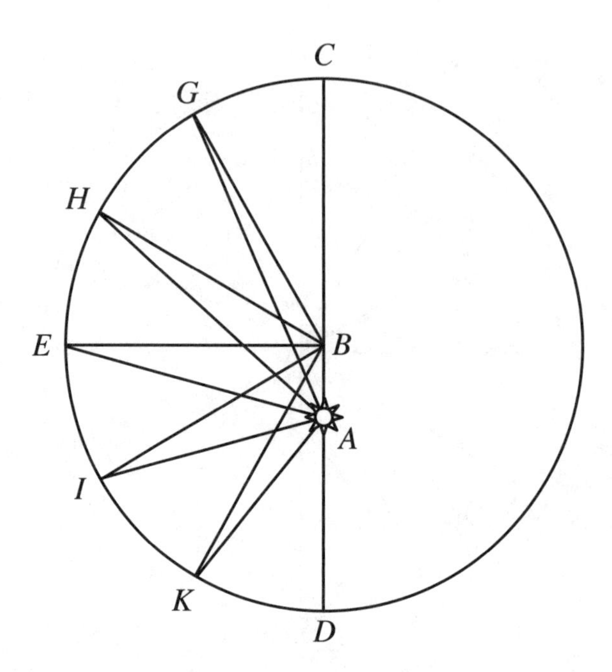

be traversed, but because I thought that the measure of the faculty by which the collected distances exert power for accumulating the times is contained in this area, so that we would be able to obtain it by knowing the area without an enumeration of least parts.

Therefore, from the above, as the area *CDE* is to half the periodic time, which we have proclaimed to be 180°, so are the areas *CAG*, *CAH* to the elapsed times on *CG* and *CH*. Thus the area *CGA* becomes a measure of the time or mean anomaly corresponding to the arc of the eccentric *CG*, since the mean anomaly measures the time.

Earlier, however, the part *CGB* of this area *CAG* was the measure of the eccentric anomaly, whose optical equation[3] is the angle *BGA*. Therefore, the remaining area, that of the triangle *BGA*, is the excess (for this place) of the mean anomaly over the eccentric anomaly, and the angle *BGA* of that triangle is the excess of the eccentric anomaly *CBG* over the equated anomaly *CAG*. Thus the knowledge of this one triangle provides both parts of the equation corresponding to the equated anomaly *GAC*.

[The rest of the chapter is omitted.]

---

3. For the "optical equation" or "optical part of the equation," see the Glossary and Appendix A.

Chapter 44
The path of the planet through the aethereal air is not a circle, not even with respect to the first inequality alone, even if you mentally remove the Brahean and Ptolemaic complex of spirals resulting from the second inequality in those two authors.

With the eccentricity and the ratio of the orbs established with the utmost certainty, it must appear strange to an astronomer that there remains yet another impediment in the way of astronomy's triumph. And me, Lord knows!—I had triumphed for two full years.[1] Nevertheless, by a comparison of the things which have been established in Chapters 41, 42, and 43, preceding, it will easily be apparent what we are still lacking.

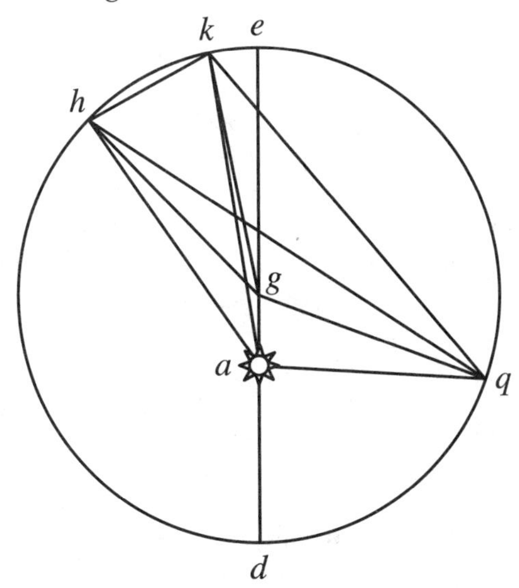

At this point, Kepler brings back the above diagram, which he had first introduced in Chapter 41. There, he had used it to determine *ag* (Mars's

1. Soon after arriving at Brahe's research center in Prague, early in 1600, Kepler was using the triangulation method to establish Mars's orbital parameters. In April of 1602, he realized that the orbit could not be circular, and that he therefore had a lot more work to do before his triumph could be celebrated.

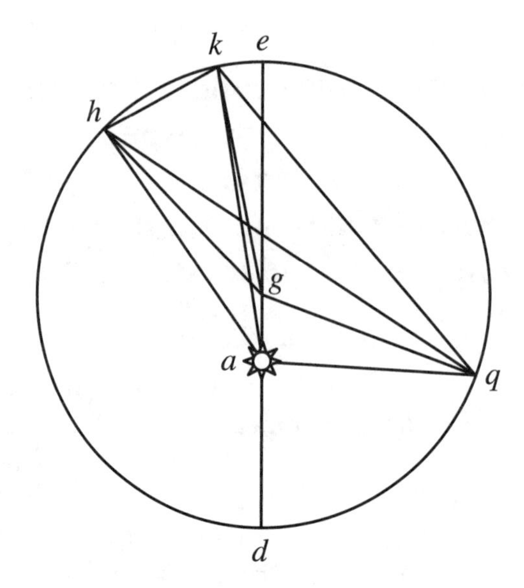

eccentricity) and the apsides *de.* Here his purpose is quite different. Using the triangulation method presented in Chapter 24, Kepler ingeniously inverts the procedure to find Mars's distances at a number of places *h*, *k*, and *q*, on the orbit. He then compares these with the distances as they would be in a circular orbit.

The determination of the distances *ah, ak,* and *aq* was very carefully carried out in Chapters 26–28 in an operation of great subtlety. A detailed presentation of these computations is beyond the scope of this module; however, the principles involved are the same as those used in Chapter 24.

Although we will take those distances as given, it is not too difficult to calculate the distances that would result from a circular orbit. Let's take as an example position *q,* which is the position of Mars that was used in Chapter 24. Again using the triangulation method, Kepler found (partly by trial and error) that the best value for Mars's heliocentric longitude (the direction of line *aq*) is 14° 21' 7" Taurus. At the same time, the longitudinal position of the aphelion, *ae,* was 28° 41' 40" Leo (from Chapter 42, adjusted to the date of the observation). Therefore, angle *eaq* or *gaq* is 104° 20' 33". The eccentricity *ag* and the diameter *gq* of the eccentric were found (by measurements with Mars near *e* and *d*) to be 14,140 and 152,640, respectively (where the radius of the earth's orbit is taken as 100,000). By the law of sines,

$$\frac{\sin aqg}{ag} = \frac{\sin gaq}{gq} = \frac{\sin agq \text{ (or } \sin egq)}{aq},$$

and therefore angle *aqg* is 5° 8' 57". But by Euclid, *Elements* I.32, *egq* = *gaq* + *aqg*, and therefore *egq* = 109° 29' 30". Substituting this into the right-hand equality of the law of sines, above, gives 148,521 as the length of *aq*, insignificantly different from Kepler's figure, 148,539.

For the other two positions, the longitude of *k* is 8° 19' 4" Virgo, and that of *h* is 5° 24' 21" Libra. Other magnitudes are the same, and the computation proceeds similarly.

Interestingly, Kepler's working notes show that when he first made this comparison, he was sure there must be some error and made a note that he must give some thought to how to adjust the planetary positions to make the orbit circular. Several weeks later, when he was comparing the area law with an equant-based theory, he realized that his *physical* theory demanded an oval orbit. Only then did he trust the observational evidence!

	ak	*ah*	*aq*
The distances on the circle come out thus:	166,605	163,883	148,539
But in the observations they were found to be:	166,255	163,100	147,750
Difference:	350	783	789

If anyone wishes to attribute this difference to the slippery luck of observing, he must surely not have felt nor paid attention to the force of the demonstrations used hitherto, and will be shamelessly imputing to me the vilest fraud in corrupting the observations of Brahe. I therefore appeal to the observations of subsequent years, at least those made by experienced observers. For if in any respect I have given free rein to my inclinations in one direction, I only go so much the farther into error on the other side. But there is no need of this. I am addressing this to you who are experienced in matters astronomical, who know that in astronomy there is no tolerance for the sophistical loopholes that beset other disciplines. To you I appeal. You see at *k* a small defect from the circle; at *h*, *q* on both sides, a rather large one, enough so that we cannot excuse it by uncertainties in observing (for in Chapter 42 I reckon an uncertainty of perhaps 200, or at most 300 units).

What, then, is to be said? Could this be the situation described in Chapter 6 above, in which by transposition of the reference point from the sun's mean motion to its apparent motion I set up another eccentric that makes an excursion towards the side of the sun's apogee? By no means. For in that case, it would approach from the one side by the same amount as it moves away on the other. Here, however, you see that the planet approaches within the circular orbit on both sides. This is confirmed by many other observations, some of which follow below in Chapters 51 and 53.

Clearly, then, [what is to be said] is this: the orbit of the planet is not a circle, but comes in gradually on both sides and returns again to the circle's distance at perigee. They are accustomed to call the shape of this sort of path "oval."

This same thing is also proved from Chapter 43 preceding. There it was supposed that the plane area of a perfect eccentric [circle] is approximately equivalent to all the distances of the equal parts of the circumference of that eccentric from the source of the motive power, however many they are. Thus, the parts of the area measure the amounts of time that the planet spends on the parts of the corresponding eccentric circumference. But if that area about which the planet marks a boundary is not a perfect circle, but is diminished at the sides from the amplitude it has at the apsides, and if nevertheless this area circumscribed by an irregular orbit still measures the times that the planet takes to traverse the whole and its equal parts, then this diminished area measures a time equal to that measured by the previous undiminished area. So the parts of the diminished area nearest aphelion and perihelion measure a greater time, because in those regions the diminution is narrowest, but the parts at the middle distances measure less time than before, because the greatest diminution in the whole area occurs there. So if we now use the diminished area in adjusting the equations, the planet will become slower near aphelion and perihelion than it was in the previous faulty form of equation, and swifter near the middle distances, because here the distances are lessened. Therefore, the times, when they are abstracted from the area and adjusted upward and downward, will be accumulated at aphelion and perihelion in much the same manner as, if one were to squeeze a fat-bellied sausage at its middle, he would squeeze and squash the ground meat, with which it is stuffed, outwards from the belly towards the two ends, emerging above and below from beneath his hand.

And indeed, if contraries remedy one another, this is plainly the most apt medicine for purging the faults under which, in Chapter 43 above, our physical hypothesis was perceived to be laboring. For the planet is going to be swifter at the middle distances, where previously it was perceived to be going slower than it should, and it will be slowed down above and below, near the apsides, where previously it did violence to the equations belonging to the eighths of the period through its excessive fleetness.

This, then, is the other argument by which it is proved that the orbit of the planet really is deflected from the established circle, making ingress towards the sides and the center of the eccentric.

But for all that, this argument still did not have enough effect upon me to let me go beyond it and think about the planet's departure from the orbit. When I had sweated for the longest time trying to reconcile equations of this sort, I was finally discouraged by the absurdity of the measurements, and abandoned the whole enquiry until later, when the distances, found in the way shown in Chapter 41, informed me about the departure from the orbit, and I once more took up this problem of the equations.

And from this, what I promised I would prove, in Chapters 20 and 23 above, is now done: that the orbit of the planet is not a circle but of an oval shape.

Chapter 57
By what natural principles the planet may be made to reciprocate as if on the diameter of an epicycle

In the preceding chapter (Chapter 56), Kepler discovered a new way of generating the distances. He had been considering the breadth of the crescent-shaped space between the circle and the true orbit, a space which he called the "lunule." He wrote,

> By two arguments, by no means obscure,...I concluded that the breadth of the lunule is to be taken as...429 units [where the circle's radius is 100,000]. While I was anxiously turning this thought over in my mind,...quite by chance I hit upon the secant of the angle 5° 18', which is the measure of he greatest optical equation. And when I saw that this was 100,429, it was as if I were awakened from sleep to see a new light.

The "new light" he saw showed him that the distances had to be shortened by the geometrically simple device of projecting the radius of the circular orbit (with one end fixed at the sun) onto a line drawn through the center of the orbit perpendicular to the line of apsides. He describes this in its epicyclic form below.

It appears, then, from the most reliable observations, that the course of the planet through the aethereal air is not a circle, but an oval figure, and that it reciprocates on the diameter of a small circle in the following manner.

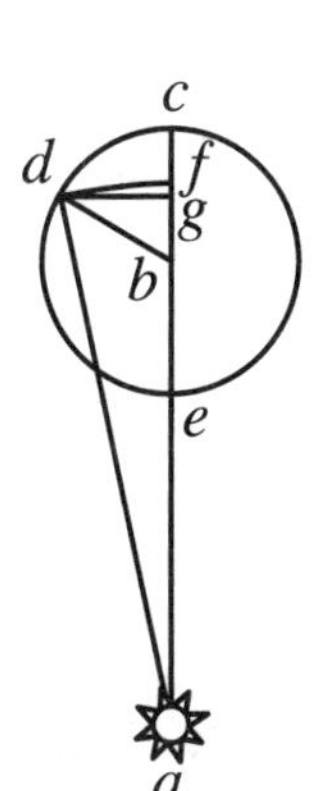

Again, Kepler brings in his imaginary epicycle *cde*, but now the distances, instead of being measured by the line *da* (or *df* which is equal to *da*), are now measured by *ga*, determined by drawing a perpendicular from *d* to *ac*. The angle *dab* is equal to the optical equation, and it reaches a maximum of 5° 18' when the angle at *b* is right. At that position, *fg* is 429 units where *ab* is 100,000.

The arc *cd* on the epicycle is proportional to the eccentric anomaly (see Ch. 39, pp. 74–5). This anomaly is measured out on the circle that coincides with the true orbit at aphelion and perihelion, but stands outside the orbit at other places. The area swept out by a line between the sun and a moving point on the circumference of this eccentric circle is the measure of the time, or mean anomaly. Even after finding the elliptical form of the orbit, Kepler continued to measure the time and the eccentric anomaly on this accessory circle, because it was much more convenient for computation.

It may appear odd that Kepler is making use of imaginary circles in constructing the orbit. However, his intent, as is made clear in the present chapter, is to identify a plausible physical process that could make the planet move in the way described by the geometrical models.

The manner in which the ascent is thus imperceptibly changed into descent by continuous diminution is more probable than if the planet were said suddenly to turn its prow in the other direction—as we have indeed said in Chapter 39, in clearly showing this to conflict with the experience embodied in the observations. And since the finger points to a natural way of measuring this reciprocation, its cause will also be natural; that is, it will be some natural—or better, corporeal—faculty, and not a planetary mind.

Also, in Chapter 39, for the best reasons, one of our suppositions was that a planet cannot make a transition from place to place by the bare effort of its inherent forces unless these be assisted or directed by an extrinsic force. If this conclusion still stands, we must as a consequence also ascribe this reciprocation in part to the solar power. In our exertions to this end, we shall be obliged once again to take up our oars which were introduced in Chapter 38.

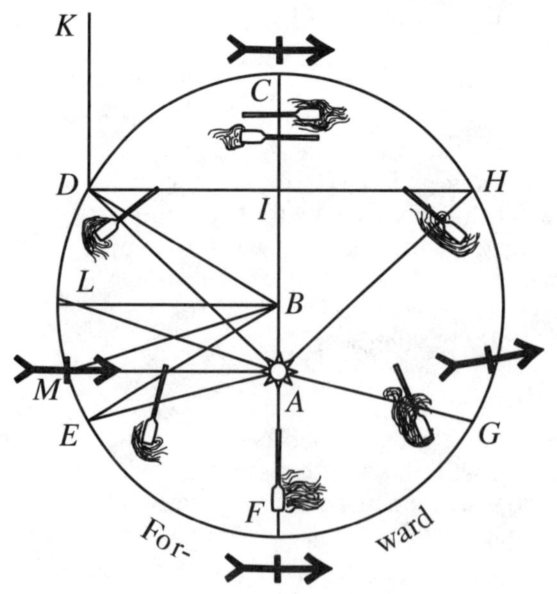

Let there be a circular river *CDE, FGH,* [flowing counterclockwise,] and in it a sailor who revolves his oar once in twice the periodic time of the planet, by an inherent and perfectly uniform force. Thus at *C* let the line of the oar be at right angles to the line from the sun, and at alternate returns let the bow and stern alternate in being forward. At *F,* however, let the line of the oar be part of the line from the sun, and at other positions let it have an intermediate inclination. Now the stream, flowing down upon the oar at *DE*, will push the ship down towards *A*, while at *C* it will push very little, since the oar is also but slightly inclined. The same is true at *F*, because at this moment the stream strikes the oar directly. At *D* and *E*, however, it pushes down more strongly, because here the oar is greatly disposed to such an approach by its inclination. The opposite happens in the ascending semicircle. For the river, coming beneath the oar at *G* and *H*, drives it away from the sun.

At the same time it will also happen, other things being equal, that the impulse will be less at *C* than at *F*, since our river is weak at *C* and strong at *F*. And this is is also in accordance with our wishes, since our reciprocation has been following equal spaces on the eccentric, and the planet spends longer in the upper ones than in the lower.

This example only shows the possibility of this arrangement. In itself it is inadequate, since the rotations of the oar and the river are accomplished, not in the same time, but a double time. Furthermore, to those looking at them from earth, the faces of the planets should appear to change, while the face of the moon, although it participates with the planets in that motion which we are discussing, does not change over the course of a month. Instead, it always is turned towards the earth, whence its eccentricity is reckoned. In addition, while the force of a river is material (for its water acts by its weight and material impetus), the force of the sun is immaterial. Therefore, the comparison with the planets ought to be different: they need no oar, no physical instrument, for catching hold of the force of some weighty thing (for that motive *species* of the sun has no weight). Nor do we deem it fitting that the stars have corporeal oars, seeing that we hold them to be round.

But from this very refutation, there comes another example, which will perhaps be more suitable.The river and the oar are of the same quality. The river is an immaterial *species* of magnetic power in the sun. So why not have the oar too borrow something from the magnet? What if all the bodies of the planets are enormous round magnets? Of the earth (one of the planets, for Copernicus) there is no doubt. William Gilbert has proved it.

But to describe this power more plainly, the planet's globe has two poles, of which one seeks out the sun, and the other flees the sun. So let us imagine an axis of this sort, using a magnetic strip, and let its point seek the sun. But despite its sun-seeking magnetic nature, let it remain ever parallel to itself in the translational motion of the globe, except to the extent that over the ages it transfers the polar direction from one of the fixed stars to another, thus causing the progressive motion of the aphelion. I nevertheless admit the possibility that a mind may be needed for both of these, of such a nature as to be adequately instructed by the animate faculty for performing this motion. For this is a motion, not of the entire body from place to place (which motion was rightly ascribed in Chapter. 39 above to a motive cause inherent in the planets), but of the parts about the center of the whole, as if at rest.

Here again, in the globe of the earth there is an example of this directional property of the axis, from Copernicus. For as long as the axis of the earth, in the annual circulation of its center, remains almost perfectly equidistant from itself in all its positions, summer and winter are brought about. On the other hand, in that over a very long period it becomes inclined, the fixed stars are thought to move forward, and the equinoxes to retrogress.

Why, then, should we have doubts about attributing to all the planets, in order to save the phenomena of eccentricity, something which is thought to be in one of them (that is, the earth) because of the phenomena of the precession of the equinoxes and the sun's annual cycle of rising and falling?

Copernicus was deceived here when he thought that he needed a special principle to cause the earth to reciprocate annually from north to south and back so as to produce summer and winter, and to bring about the equality of the tropical and sidereal years (to the extent that they are equal) by its efforts at producing equal periods. For all those effects are obtained by having the earth's axis, about which the diurnal motion is made, retain a single, constant direction: there is no need for extrinsic causes, except to account only for the extremely slow precession of the equinoxes. And so here, too, there is nothing to suggest that there will be a need for movers for the planet, which would carry its body about the sun in a parallel position, and would at the same time perform the reciprocation. For the one will naturally depend upon the other. The only thing remaining to be considered is the extremely slow progression of the aphelia.

To continue: when the strip is at *C* and *F*, there is no reason why the planet should approach or recede, since it holds its ends at equal distances from the sun, and would undoubtedly turn its point towards the sun if it were allowed to do so by the force that holds its axis straight and parallel. When the planet moves away from *C*, the point approaches the sun perceptibly, and the tail end recedes. Therefore, the globe begins perceptibly to navigate towards the sun. After *F*, the tail end perceptibly approaches, and the head end recedes from the sun. Therefore, by a natural aversion, the whole globe perceptibly flees the sun. And when it is across from *A*, where the length of the axis is pointed directly at the sun, its approach in the former situation, or its flight in the latter, is strongest.

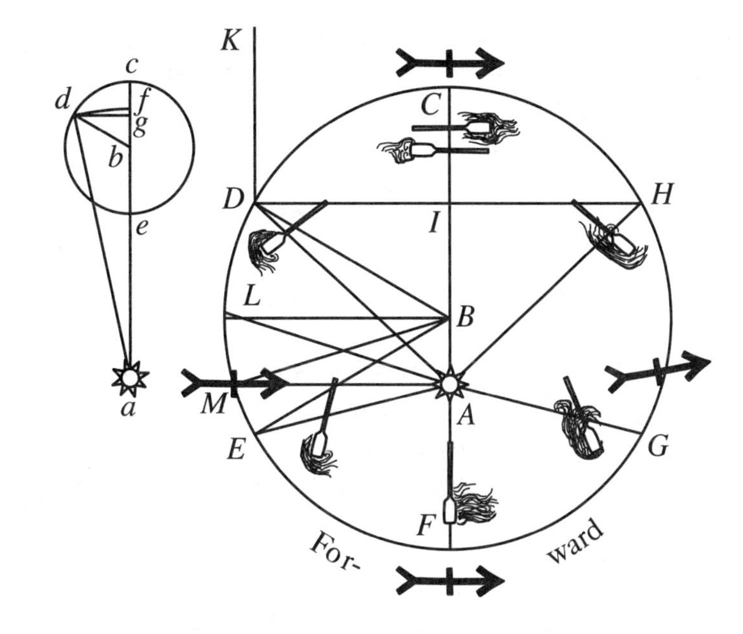

Furthermore, our earlier presuppositions derived from the observations postulated this. For of the parts of the reciprocation [on the small epicycle] which correspond to equal arcs on the eccentric, the parts at the middle were longest, and those near *c* and *e* were short.

...

In fact, it is possible for us to transfer that force which keeps the magnetic axis in a parallel position, and does not allow it to remain pointed towards the sun, from the occupation of a mind, to which we had entrusted it a little earlier, to the function of nature. It would appear to be an objection to this, that nature always acts in one and the same way, while this retentive force appears to make its exertions differently at different times. This is seen, for example, in the tendency of the axis to incline towards the sun, for the impeding of which the retentive force is ordained, which tendency is evanescent at the middle distances but most strongly evident at aphelion and perihelion. Nevertheless, what is there to prevent this force of retention's being in many places stronger than the tendency to incline towards the sun, so that the force is either not at all or but little wearied by such a weak adversary? Let us again take an example from the magnet. In it are manifestly mingled two powers, one of directing it towards the pole, and the other of seeking iron. Thus if a strip or nautical needle be directed towards the pole, while some iron approach from the side, the needle gradually would decline from the pole and incline towards the iron, thus indulging somewhat in its intimacy with the iron, but in such a way that it gives most of it to the pole. Indeed, Gilbert thinks this to be the reason why a strip declines from the pole towards the continents of greatest magnitude, the cause thus lying in the tracts of land, being greater and having a more vigorous power on the right or left to the extent that they are higher in that vicinity.

...

I will be satisfied if this magnetic example demonstrates the general possibility of the proposed mechanism. Concerning its details, however, I have doubts. For when the earth is in question, it is certain that its axis, whose constant and parallel direction brings about the year's seasons at the cardinal points, is not well suited to bringing about this reciprocation or this aphelion. The sun's apogee, or earth's aphelion, today closely coincides with the solstitial points, and not with the equinoctial, which would fit our theory; nor will it have remained at a constant distance from the cardinal points. And if this axis is unsuitable, it seems that there is none suitable in the earth's entire body, since there is no part of it that rests in one position while the whole body of the globe revolves in a ceaseless daily whirl about that axis.

So indeed, there may be absolutely no material, magnetic faculty that can accomplish the tasks entrusted to the planets individually, since there may be a lack of means, that is, no suitable diameter of the body that remains parallel to itself as the body is moved around. For this lack has just been made apparent in one of the planets, namely, the globe of the earth. Therefore, a mind must be summoned, which, as was said in Chapter 39, arrives at a knowledge of the distances it assumes by contemplating the growth of the sun's diameter. This mind would need to govern a faculty, either animate or natural, that keeps its globe in a parallel position in a manner allowing it to be suitably impelled by the solar power and to reciprocate with respect to the sun. (For a mere mind, unassisted by faculties of a lower order, cannot by itself do anything in a body) At the same time care should be taken that the periodic time of the reciprocation not be made exactly equal to the periodic return of the planet, so that the apsides will move. The plausibility of these things is argued in Chapter 39 above.

The rest of Chapter 57, concerned with how precisely the moving power can satisfy the quantitative requirements of the reciprocation, is omitted.

Editor's Epilogue: What about the ellipse?

Astronomia Nova is perhaps best known for its introduction of elliptical orbits into astronomy. It may therefore seem surprising that the ellipse itself is not prominently featured as the goal towards which we have been working. However, we must keep in mind that Kepler was attempting something much more radical than merely replacing one geometrical model with another: his aim was to reestablish astronomy on a sound physical basis. In a world of pushes and pulls, rivers and magnets, one should not expect to find pure geometrical forms.

Yet a form did emerge, to Kepler's evident surprise—he spent the better part of Chapter 59 trying to explain how the mutual tempering of two independent powers could result in a tidy geometrical form and not a new "pretzel" (see Chapter 1, p. 34). Although it didn't have the perfect simplicity of the circle, it could have been a lot messier.

But, to go back to the story where we left off, in Chapters 56 and 57, Kepler had figured out how to relate the planetary distances to the (by now purely fictitious) eccentric anomalies. And since the area swept out on the eccentric represents the time, Kepler could relate the distances to the times. The remaining problem was to find out how to relate these distances and times to the observed positions.

Kepler didn't have much to guide him, other than the observations. What he discovered, by trial and error, was that the points on the eccentric circle had to be squeezed perpendicularly inward, towards the line of apsides. In the diagram, point *L* was substituted for *G*, *M* for *H*, and so on. Kepler knew from Commandino's commentary on Archimedes' *On Conoids and Spheroids* that the curve *CLMFNOD* would be an ellipse. He was then able to work out a series of original geometrical proofs showing that the distances *AL*, *AM*, and so on, are the same as

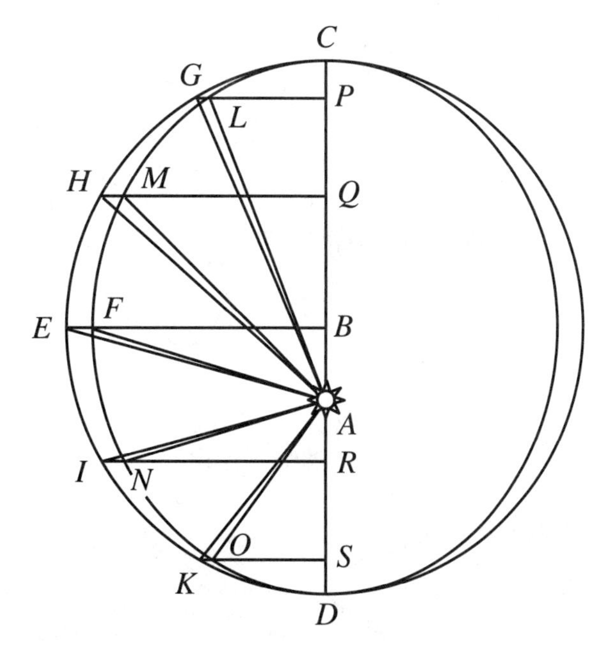

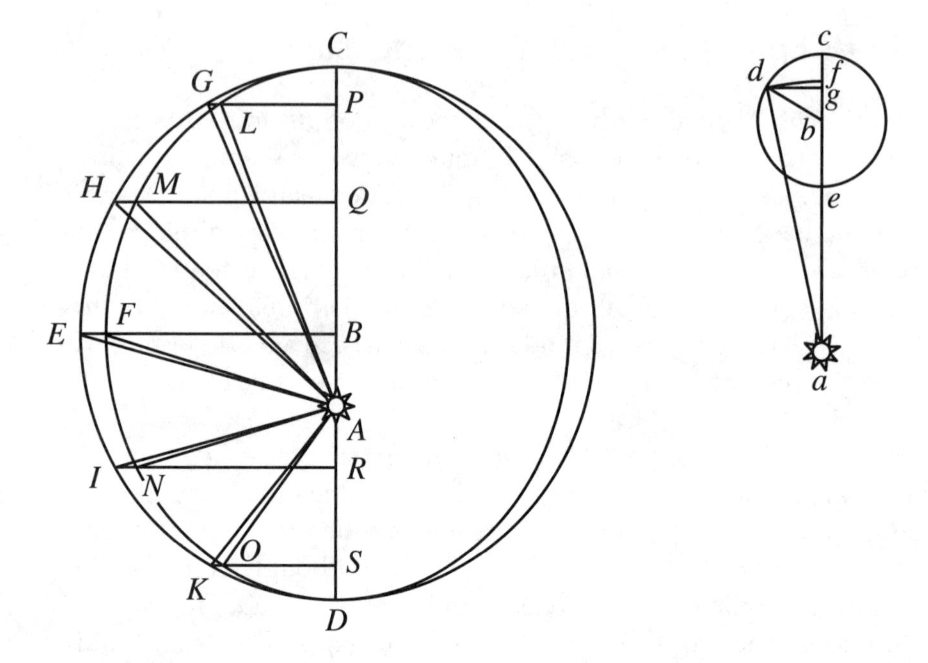

the distances of the reciprocation on the imaginary epicycle. That is, if arc *cd* is made equal to arc *CH*, then *AM* will be equal to *ag*. He also showed, using the observed positions of Mars, that if the areas *ACL, ACM,* and so on (or, equivalently, areas *ACG, ACH,* and so on) are made proportional to the time, the planet will lie somewhere on the lines *AL, AM,* and so on, at those times. Combining these two discoveries, he concluded that the planet must lie exactly at the points *L, M, F,* and so on—that is, on an ellipse.

Curiously, although Kepler was familiar with the focal properties of the ellipse, he seems to have been unaware at the time of writing the *Astronomia Nova* that the point *A*, the sun's position in the above diagram, is a focus. That fact was made explicit only later, in his *Epitome of Copernican Astronomy* (1618–1621).

Appendix A
Geometrical planetary models of Kepler's time

Although the selections in this module are comparatively nontechnical, Kepler assumes familiarity with the general features of the geometrical model of planetary motion that had been presented by the Greek astronomer Claudius Ptolemy (second century CE). What follows is a very cursory sketch of the Ptolemaic models. A much more thorough and informative account of them, together with an explanation of how they relate to Copernicus's theory, is provided by Michael J. Crowe, *Theories of the World* (See the Bibliographical Note for a complete citation).

The most striking feature of Ptolemy's model is the epicycle, which generates the series of retrograde loops that Kepler illustrates so tellingly in Chapter 1 of *Astronomia Nova*. The epicyclic model is depicted in simplified form in the first diagram to the left.

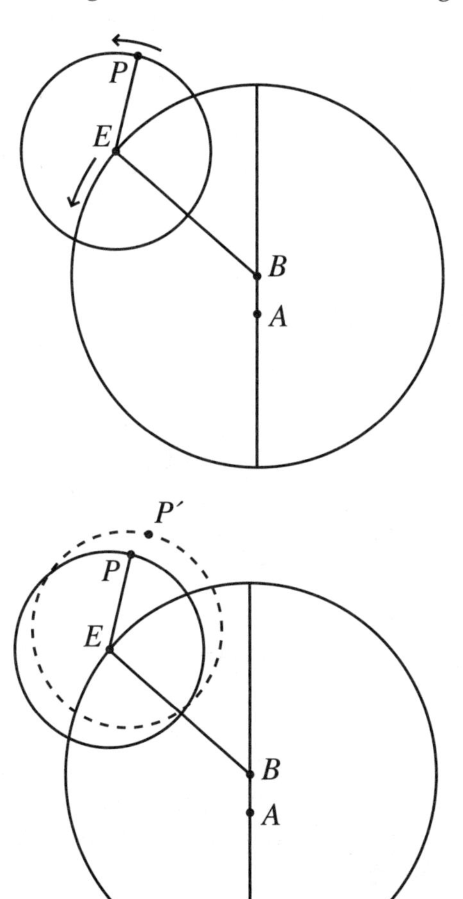

The planet, P, moves uniformly along the circumference of a circle (the **epicycle**) with center E. The point E simultaneously moves in the same direction on a larger circle (the **deferent** or **eccentric**), whose center B is removed by some distance AB (the **eccentricity**) from the earth A.

This model didn't quite work, however: the planet was seen, not at P, but higher up, at P', in the second diagram. To enlarge the epicycle or increase the eccentricity of the deferent

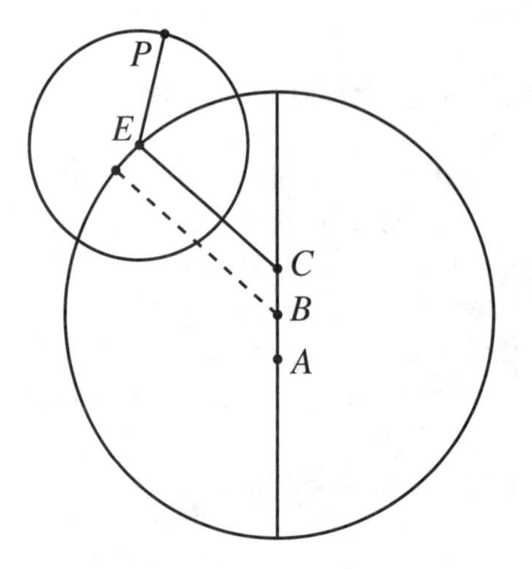

wouldn't work, so Ptolemy adopted the expedient of changing the center around which the center of the epicycle *E* moves uniformly. This new point is called the **equant** (*C* in the adjacent diagram). In Ptolemy's models, $BC = AB$; that is, the **eccentricity of the equant** *AC* is exactly twice as great as the **eccentricity of the eccentric** *AB*. This is called **bisection of the eccentricity**.

Kepler, as a follower of Copernicus, realized that the epicycle is really an artifact of the earth's motion, an image of the earth's orbit superimposed on the planet's own orbit. Using his triangulation method exemplified in this module by Chapter 24, he was able to show that in Ptolemy's theories, no less than in Copernicus's, the motion of the earth or sun had to be nonuniform. What this meant for Ptolemy (as he points out on p. 7 of the Introduction) is that each epicycle needs its own equant (in addition to the one for the eccentric) in order to make the motion of *P* nonuniform on its circle. That is, there must be one center *E* around which the planet moves uniformly, but another point nearby, moving around the deferent with *E*, from which *P* maintains a constant distance.

What this showed very clearly was that it doesn't make sense to try to include the theory of the earth (or sun), with all its complication, in the same diagram along with that of Mars. Instead, it is much neater to separate out the epicycle, by treating it as the motion of the observer on earth, in a separate computation. Kepler therefore treated the theories in this way even when he was considering the traditional Ptolemaic models. So when he refers to the Ptolemaic form of hypothesis, he is thinking of the geometrical model without the epicycle, as depicted on the opposite page.

Once the epicycle is separated out, this model will work equally well in both geocentric and heliocentric systems. *A* is the earth (geocentric) or sun (heliocentric), *D* and *E* are respectively **apogee** and **perigee** (geocentric) or **aphelion** and **perihelion** (heliocentric). *DE* is the **line of apsides**. *P* is the planet (heliocentric) or the center of the epicycle (geocentric).

There are five important angles in this figure: three angles of **anomaly** and two **equations**. The anomalies are all angles around the three centers. They have that name for historical reasons and not because they are abnormal or irregular in the usual sense of the word "anomaly." The equations are the differences between the anomalies.

The angle *DCP* is the **mean anomaly**. It is proportional to the time, starting from aphelion; in Mars, it increases by a little over half a degree per day.

The angle *DBP* is the **eccentric anomaly**. It is a measure of the length of the arc *DP* that the planet (or the center of the epicycle) traverses on the eccentric.

The angle *DAP* is the **equated anomaly**. It is the observed position of the planet (or center of the epicycle) as seen from the sun (or earth).

The total difference between the mean anomaly and the equated anomaly is called the **total** or **whole equation**. It is the angle *APC*. The equation has two parts, angles *APB* and *BPC*, each of which is computed separately. Kepler calls the angle *APB* the **optical equation**, because it is the purely optical effect of moving the observer from the center of the eccentric *B* to point *A*. He calls angle *BPC* the **physical equation**, because it represents a real, physical speeding up and slowing down of the planet (or the epicycle) on the eccentric.

In the course of the readings in this module, some other terms will unavoidably be used that come ultimately from ancient planetary theory. These terms are not complicated, but they are unfamiliar to modern readers. A brief explanation of these terms is therefore provided in the Glossary.

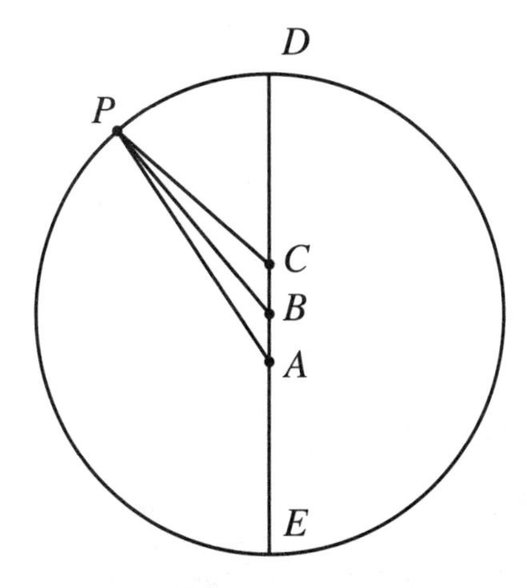

Appendix B
Technical information useful for a closer reading

"Supposing the same planet to be in turn at two distances from the sun, remaining there for one whole circuit, the periodic times will be in the duplicate ratio of the distances or magnitudes of the circle," (Chapter 39, p. 71, at footnote 1).

This is a consequence of the demonstrations in Chapters 32 and 33. It can be demonstrated as follows.

The periodic time over a circumference is equal to the arc length of the circumference divided by the speed:

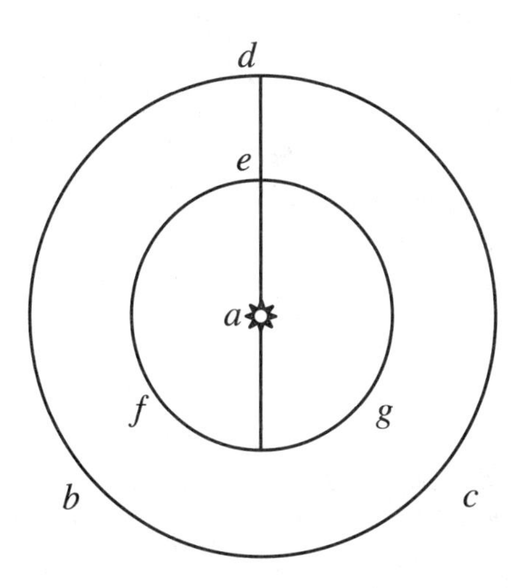

T_{dbc} = circ. *dbc* / speed *d*, and

T_{efg} = circ. *efg* / speed *e*.

Turning this into a proportion,

$$\frac{T_{dbc}}{T_{efg}} = \frac{\left(\dfrac{\text{circ. } dbc}{\text{speed } d}\right)}{\left(\dfrac{\text{circ. } efg}{\text{speed } e}\right)}$$

or

$$\frac{T_{dbc}}{T_{efg}} = \left(\frac{\text{circ. } dbc}{\text{circ. } efg}\right)\left(\frac{\text{speed } e}{\text{speed } d}\right).$$

But the circumferences are equal to 2π times the respective radii *ad* and *ae*, and by Chapter 32, the speeds are inversely proportional to the radii.

Therefore,

$$\frac{T_{dbc}}{T_{efg}} = \left(\frac{2\pi\, ad}{2\pi\, ae}\right)\left(\frac{ad}{ae}\right) = \frac{ad^2}{ae^2}.$$

The ratio of the squares of the radii is the same as the duplicate ratio of the radii (see Euclid, Elements, V def. 9 and VI.20 with its porism), and therefore the periodic times are in the duplicate ratio of the radii or distances of the planet from the sun.

"Thus we humans know that the sun is 229 of its own semidiameters distant from us when its diameter subtends 30′, and 222 semidiameters when it subtends 31′," (Chapter 39, p. 74, at footnote 4).

Suppose that the sun, QSR, as seen from the earth T, subtends 30' (that is, that the angle QTR is 30').

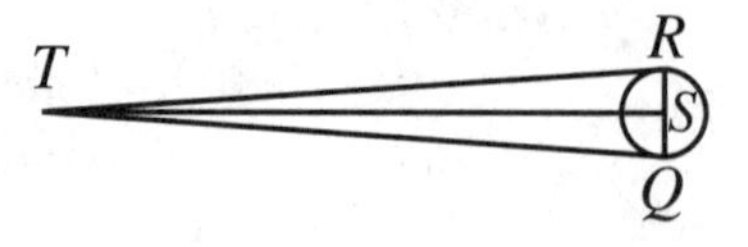

Then angle STR is 15', and, by definition of the sine, SR/TR is the sine of this angle, 0.004363.

The inverse of this is about 229, which is the number of times the semidiameter SR measures the distance TS or TR, which are practically equal.

This is insignificantly different from the number given by Kepler. The case in which QSR subtends 31' is solved similarly.

"The motive mind will have to have a very good memory in order to adjust the unequal versed sines of the arcs on the epicycle to the equal increases of the solar diameter," (Chapter 39, p. 76, at footnote 6).

The versed sine of an angle is the difference between the radius of the reference circle and the cosine of the angle. In the adjacent diagram, if a perpendicular dg is dropped from d to bc, the versed sine of angle dbc, or of arc dc, is cg. However, in the previous paragraph, Kepler wrote that the sun's apparent diameter is to be made inversely proportional to the distances. This is shown as follows.

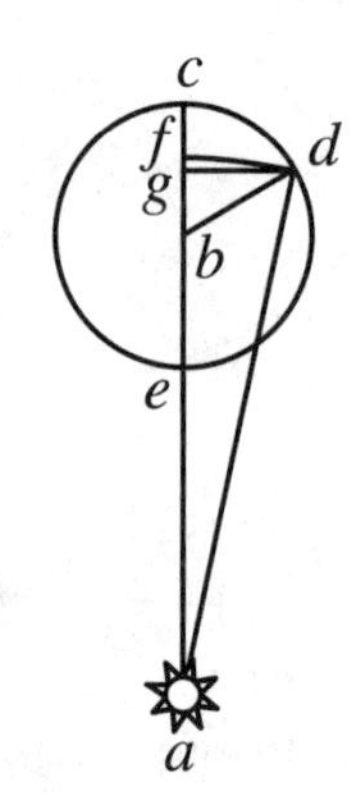

From the diagram above, by definition,

$\sin STR = \frac{1}{2}QR \,/\, ST$, since $ST = RT$, very nearly,

and because angle STR is small, the angle and its sine are nearly proportional, so that

angle $STR \propto \frac{1}{2}QR \,/\, ST$, approximately; that is (multiplying both sides by 2),

angle $QTR \propto QR \,/\, ST$

or since $\frac{1}{2}QR$ is constant, the sun's apparent diameter is approximately inversely proportional to the distances. But the distance

ST is the same as distance *da* or *fa* in the epicycle diagram above, and on the assumption of a circular orbit this distance corresponds to an eccentric anomaly that is here represented by the arc *cd* (see the note on p. 72, above). So, as the planet moves over an eccentric arc *cd*, it moves inward from its maximum distance *ca* to a new distance *cf*, and the distance *cf* is obviously not proportional to the arc *cd*, since it is approximately equal to the versed sine *cg*, which is $bc - bd \cos dbc$. Kepler is evidently treating *cg* and *cf* as approximately equal, though (as will become evident in Chapter 57) the difference is crucially important.

Glossary

The following are technical terms that are used by Kepler in the present selection of readings from *Astronomia Nova.* A brief exposition of the fundamental planetary model referred to by Kepler (following Ptolemy) is included in Appendix A. Since that account shows the way in which many of these terms were used in context, readers are encouraged to consult the appendix first, using this Glossary as a supplementary tool.

Words that are in bold type (other than the lead words of the individual entries) are terms defined elsewhere in the Glossary.

Acronychal: a word taken over from Ptolemy's Greek that literally means "night rising." It refers to the position of a planet that is rising at the same time the sun sets; that is, when the planet is at opposition to the sun.

Anomaly: Anomaly in general is a measure of angular position relative to **aphelion** (or **apogee**, in an earth-centered theory). There are three kinds of anomaly, distinguished by the centers around which the angular positions are measured. The **mean anomaly** is measured around the **equant**, and is proportional to the time. The **eccentric anomaly** is measured around the center of the **eccentric**. The **equated anomaly** is measured around the center of the system (the sun, for Copernicans; the earth, for Ptolemy). For an account of the general geometrical planetary model, see Appendix A.

Aphelion: in a heliocentric model, the point on the **eccentric** farthest from the sun.

Apogee: in a geocentric model, the point on the **deferent** (or **eccentric**) farthest from the earth.

Apsis (plural, **apsides**): the collective name for **aphelion** and **perihelion** (in a sun-centered system), or **apogee** and **perigee** (in an earth-centered system). The **line of apsides** is the line joining the pair of apsides.

Deferent: In the Ptolemaic model, the circle on which the **epicycle** rides. In the model that Kepler calls "Ptolemaic" (see Appendix A), there may or may not be an **epicycle**, depending on whether the context is sun-centered or earth-centered. Therefore, Kepler usually calls this circle the **eccentric**, because it is always eccentric with respect to the center of whatever system it is in.

Diurnal: taking place over the course of a day.

Eccentric: see **deferent.**

Eccentric anomaly: see **anomaly**.

Ecliptic: the projection of the earth's (or sun's) orbit onto the celestial sphere.

Epicycle: a circle whose center is carried around on another circle, called the **deferent**. Usually, a planet moves around the circumference of the epicycle, but sometimes (as in Copernicus's models) an epicycle acts as a deferent for another epicycle.

Equant or **equalizing point**: a point about which the angular motion of a planet (or of the center of the **epicycle**) is uniform.

Equated anomaly: see **anomaly**.

Equation: the angular correction that must be added to or subtracted from a planet's **mean anomaly** to give its **equated anomaly**. The equation has two parts: the **optical equation**, which is the effect of displacing the point of observation (the center of the system) from the center of the **eccentric**; and the **physical equation**, which results from the real speeding up and slowing down of the planet, produced by the **equant**.

First inequality, see **Inequality**.

Inequality: apparent changes in speed and direction of a planet's motion. Kepler followed the ancients in distinguishing two kinds of inequalities. The **first inequality** (which Ptolemy called the "zodiacal anomaly") is a planet's tendency to move faster in one region of the sky and slower at the opposite region (the effect of the second inequality being ignored). The **second inequality** (which Ptolemy called the "heliacal anomaly") is the tendency of a planet to move swiftly and in a forward direction when in conjunction with the sun (or at superior conjunction, for Venus and Mercury), and slowly or even backwards when opposite the sun (or at inferior conjunction, for Venus and Mercury). See Chapter 1 for Kepler's account of these inequalities.

Latitude: the angular distance of a planet north or south of the **ecliptic.** It is the apparent effect of the planet's orbit lying in a plane that is somewhat inclined to the plane of the earth's orbit. Very early in his study of Mars, Kepler realized that once the mutual inclination of the two orbits was established, the apparent latitude of Mars could be used to find its distance from the sun in relation to the size of the earth's orbit.

Line of apsides: see **apsis**.

Longitude: the position of a planet on the **ecliptic**, measured from the spring equinox point. There are two kinds of longitude: the **mean longitude**, which, like the **mean anomaly**, is proportional to the time, and the **true longitude**, which gives the position of the planet itself with respect to the ecliptic.

There is also another sense in which the Latin word "longitudo" is used: as a component of the (always plural) term *longitudines mediae*. There is considerable textual evidence that Kepler meant "lengths" or "distances" here, rather than angular positions. This phrase has therefore been translated "middle distances." It refers to the region where the distances from a planet to the sun are intermediate between the **aphelial** distance and the **perihelial** distance.

Mean anomaly: see **anomaly**.

Middle distances: *longitudines mediae*. See the explanation under **longitude**.

Optical equation: see **equation**.

Perigee: the point of a planet's nearest approach to the earth.

Perihelion: the point of a planet's nearest approach to the sun.

Physical equation: see **equation**.

Primum mobile: in earth-centered systems, the outermost sphere, whose period of rotation is slightly less than one day.

Retrograde: When Mars or another of the outer planets is **acronychal**, after passing the point of opposition to the sun it ceases its forward motion along the zodiac, briefly becomes **stationary**, moves backward through the point of opposition, stops again, and finally moves forward, once again passing the point of opposition and continuing on its way. This reversal is called "retrograde motion," and is a necessary consequence of the Copernican arrangement of the planets (for Ptolemy, it was a curious accident). For a clear account of this, see Crowe, *Theories of the World*, pp. 92–4.

Second inequality, see **Inequality**.

Species: This Latin word, related to the verb *specio* (see, observe), has an extraordinarily wide range of meaning. Its root meaning is "something presented to view," but it can also mean "appearance," "surface," "form," "semblance," "mental image," "sort," "nature," or "archetype," to mention only a selection of its most diverse senses. It is in fact the Latin equivalent of the Greek **εἶδος**, which is Plato's word for his "forms" or "ideas."

Because *species* is an important term in Kepler's physics, a consistent translation was needed. Yet, after consideration of many possible candidates, no single English word appeared capable of carrying the range of meanings without seriously misleading the reader. The translator, in the end, has thrown up his hands, admitted defeat, and has declined to translate it at all. It appears as the Latin word "*species*," always in italics, and is pronounced "SPECK-ee-ace," to distinguish it from its English cognate.

Stationary, see **retrograde**.

Bibliographical Note

The selections translated here originally appeared in the editor's complete translation of *Astronomia Nova,* published by Cambridge University Press in 1992 (now out of print). In 2015, Green Lion Press published a new edition in a larger format with many corrections, appendices on Kepler's use of Brahe's observations, and a new index.

All of Kepler's published works and surviving correspondence, and a selection of manuscripts, have been published in the original languages in a fine modern edition: *Johannes Kepler Gesammelte Werke,* which began to appear in 1938 and is now complete (22 volumes in 25). It is published by C. H. Beck in Munich. *Astronomia Nova* is in Volume 3. It is cited as *KGW* in the present selections.

Citations of Kepler's *Optics* are from the editor's translation (Green Lion Press, 2000). The original Latin edition, published in 1604, bore the title *Ad Vitellionem paralipomena, quibus Astronomiae pars optica traditur.* Kepler sometimes refers to it as *Optica* and sometimes as *Astronomiae optica.* The Latin text of the *Optics* is in Volume 2 of *KGW.*

In Chapter 7, Kepler refers the reader to a book by Georg Joachim Rheticus, entitled *Narratio Prima,* or "First Account." This was the first published account of the Copernican system, and contains (along with other more abstruse matters) a very clear description of how the sun-centered system works. It has been translated by Edward Rosen, and is included in Rosen's *Three Copernican Treatises* (New York: Octagon Books, 1971 and other editions).

There are a number of excellent studies of *Astronomia Nova.* Bruce Stephenson, *Kepler's Physical Astronomy* (Princeton University Press, 1994) is an in-depth study of the text, and also discusses Kepler's other chief astronomical works. James R. Voelkel, *The Composition of Kepler's Astronomia Nova* (Princeton University Press, 2001), shows how the book came to be written, and why it took the unusual form in which it appears. Rhonda Martens, *Kepler's Philosophy and the New Astronomy* (Princeton University Press, 2000), considers the philosophical aspects of Kepler's astronomy.

Further background reading on the development of astronomy before Kepler is provided by Michael J. Crowe, *Theories of the World from Antiquity to the Copernican Revolution* (Dover Publications, 1990). Crowe includes substantial selections from the writings of Ptolemy and Copernicus, explanations of planetary models, and much historical information.

About this Series

SCIENCE CLASSICS FOR HUMANITIES STUDIES is a series of study modules designed to bring fundamental works of science and mathematics within the grasp of students and other readers without the need for specialized preparation. The series reflects the Green Lion's conviction that an understanding of science, and especially of the classical works of science, is essential for all students of the humanities. Science, no less than poetry or philosophy, is human thought, a response both to the outer world of our senses and the inner experience of our consciousness. The more profound a scientific work is, the more directly it addresses itself to our humanity; therefore, there is much in the greatest works of science that can be grasped without special preparation. Yet too many educational programs find themselves limited by the supposed divide between the humanities and the sciences—the so-called "two cultures."

Further, teachers and institutions who wish to heal this unnecessary fracture have had to confront two discouraging barriers. On the one hand, classic texts of real science are often found to be forbiddingly technical in content and burdened with terminologies either antiquated or arcane. On the other hand, popularizations of these classics insulate students from the actual workings of thought and imagination that classic texts embody. Green Lion Press has addressed this dilemma with the series SCIENCE CLASSICS FOR HUMANITIES STUDIES, issued in slim, inexpensive student editions under the *Green Cat Books* imprint.

Each volume in the series is a compact, inexpensive presentation of classic scientific and mathematical texts, offering generous but judicious guidance for the reader. We have drawn on our many years of reading these books with nonspecialist students to choose selections of real substance, and to provide helps that make the texts accessible while at the same time allowing the original texts to speak for themselves, in their own voices.

Besides humanities students, this series will be of interest to those interested in science but lacking time or expertise to read these works unabridged and without assistance. It will also serve readers who already enjoy a technical background but who may wish to experience more directly the sources of contemporary scientific concepts.

Classic works of science and mathematics, no less than other works of literature, drama, and philosophy, lead us to questions (and answers) that may enlighten or delight us, or may lead us to a new understanding of the multiplex and often conflicting views of reality presented in great scientific works. It is the Green Lion's aim to enable readers not only to observe but to participate in such significant achievements of thought. Other volumes in the SCIENCE CLASSICS FOR HUMANITIES STUDIES series focus on Faraday's *Experimental Researches in Electricity*, Newton's *Principia,* the first Book of Euclid's *Elements*, Darwin's *On the Origin of Species,* and Lavoisier's *Elementary Treatise on Chemistry.*